LANCASTER PRESS, INC., LANCASTER, PA.

TRANSACTIONS

OF THE

AMERICAN PHILOSOPHICAL SOCIETY

HELD AT PHILADELPHIA

FOR PROMOTING USEFUL KNOWLEDGE

NEW SERIES—VOLUME XXXII, PART II

JANUARY, 1942

EGYPTIAN PLANETARY TEXTS

O. NEUGEBAUER

ON SOME ASTRONOMICAL PAPYRI AND RELATED PROBLEMS OF ANCIENT GEOGRAPHY

O. NEUGEBAUER

PHILADELPHIA:

THE AMERICAN PHILOSOPHICAL SOCIETY

104 SOUTH FIFTH STREET

1942

EGYPTIAN PLANETARY TEXTS

By O. Neugebauer

Brown University, Providence, R. I.

Table of Contents

Preface

There is so very little known about Egyptian astronomy that I do not feel obliged to explain at length why I undertook a complete publication of the only two texts which give us any information about an Egyptian description of the planetary movements, even if they belong to so late a period as Roman times. I must however express my gratitude for the assistance given me in my undertaking, first of all by the authorities of the museums in Liverpool and Berlin and of the library of the University of California in supplying excellent photographs. The main burden of the very extensive numerical calculations carried out in order to compare the positions given by the ancient texts with the actual orbits of the planets rested on the shoulders of Mr. Olaf Schmidt who has therefore a very considerable share in this edition. I feel especially grateful to the American Philosophical Society for accepting the publication of this study in its Transactions.

O. N.

Abstract

The only two Egyptian manuscripts treating the movement of the five planets are the Demotic Berlin Papyrus P.8279 ("P") and the Stobart tablets ("S") now in Liverpool. Both texts were edited previously, P by Spiegelberg, S by Brugsch, but both without detailed investigation of their astronomical content. This is undertaken in the present edition and leads to an almost complete restoration of both texts.

The text P covers the years 16 B.C. to 11 A.D., S, with larger gaps, the years Vespasianus 4 (= 71 A.D.) to Hadrian 17 (= 132 A.D.). The positions of the planets recorded, compared with modern calculations, reveal a systematic deviation in the longitudes which can be explained by the use of a fixed point as origin on the Egyptian zodiac. An extensive investigation is devoted to the Greek records on Egyptian astronomy. As a result, it follows that Clemens Alexandrinus is the only reliable source, while the other known sources follow a common pattern without historical value. The close parallelism between the Demotic texts and the Greek papyrus Tebtunis II,274 makes it very probable that these planetary texts belong to the class of so-called "eternal tables" criticized by Ptolemy. This seems to indicate that these texts are of Hellenistic origin.

The latter part of the memoir is devoted to problems of terminology and palaeography. The reason for the interchange of ordinary number signs and date number signs is set forth. Investigation of errors indicates that P was copied from a Hieratic original.

§ 1. Introduction

*With these intervals as their period,
the planets show again and again all the
phenomena which God desired to be seen
from the earth.*

RHETICUS, Narratio prima.

1. The texts treated here are the large Demotic papyrus P.8279 in the collection of the *Staatliche Museen Berlin* and the so-called "Stobart Tablets" i.e. four wooden tablets from the Rev. H. Stobart Collection now in the *Free Public Museum at Liverpool*.

Both texts have been published earlier. In 1855 there appeared a pamphlet "Egyptian Antiquities", published by Rev. H. Stobart,[1] containing five plates of very large format only. Plate II[2] is called "4 astronomical tablets on wood written in the Demotic character—found at Thebes"; the drawings have been made "under the direction of Dr. H. Brugsch." In the following year Brugsch himself published a little book under the title "Nouvelles Recherches sur la Division de l'Année des Anciens Egyptiens" containing as second part a "Mémoire sur les observations Planétaires consignées dans quatre Tablettes Egyptiennes en Ecriture Démotique", where he again reproduced the four Stobart tablets (with slight alterations in details), together with a French translation and commentary. A second text of this kind did not appear until about 50 years later, namely the Berlin Papyrus P.8279, edited by W. Spiegelberg. Spiegelberg, however, restricted himself to a German translation with some short introductory remarks and a photographic reproduction of only a few columns. A comparison of his translation with the complete text shows easily that there are many instances where the readings can be completed or gaps restored. In view of this and the mistakes in Spiegelberg's translation, a new edition seems necessary.

Brugsch's publication of the Stobart tablets is in general much more trustworthy. However, Brugsch desisted from the treatment of the largest part of one side of tablet IV, which is badly preserved and therefore not readable, before the astronomical principle in the arrangement of the numbers became fully clear. Furthermore, he went partly astray in the chronological arrangement of the tablets. Neither the Stobart tablets nor P.8279 are explicitly dated, and both only mention "years" without any more explanation. The Stobart tablets, marked by Brugsch respectively as I, II, III, IV, contain the following "years":

I: from year 9 to 15
II: from year 16 to 19 and from 1 to 3
III: from year 4 to year 10
IV: from year 11 to year 17.

From this sequence of numbers Brugsch[3] concluded that those texts belong to "les années 8 – 19 du règne d'un roi égyptien et de l'an 1 – 17 de son successeur" and he conjectured that the first was Trajan, the only Roman emperor who reigned exactly 19 years; the second must therefore be Hadrian.[4]

It is very simple, however, to show that tablet III cannot be the continuation of tablet II. It is sufficient to consider e.g. the movement of Saturn. The last position of this planet as indicated in tablet II is "year 3, month XI, zodiacal sign Cancer"; the first position of the same planet in tablet III is "year 4, month I, sign Scorpio". Now it is impossible for Saturn to move in two months from Cancer to Scorpio, which corresponds to a movement of 120°, because Saturn moves about 12° a year! By the same reasoning the arrangement III → IV, which would ascribe to Saturn a movement of 150° in one month, is impossible. Corresponding discrepancies appear for the other planets, of course less obvious the nearer the planet is to the sun. On the contrary the arrangement I → II is in perfect agreement with the astronomically expected facts. The systematic investigation of all four tablets by astronomical calculation gives the following result:[5] tablet III is the oldest, referring to the years 4 to 10 of Vespasianus; then follow I and II from Trajan 9 to Hadrian 3, confirming here Brugsch's expectation;[6] finally IV refers to the years Hadrian 11 to 17.

The gap between tablets III and I corresponds to the time covered by four tablets of the same size as the preserved tablets; the gap between II and IV can be filled by one tablet exactly. We introduce therefore the following new notation for these tablets:

A	(= Brugsch III)
B₁ to B₄	lost
C₁ and C₂	(= Brugsch I and II)
D	lost
E	(= Brugsch IV).

The complete set from A to E would have covered the years Vespasian 4 = 71 A.D. to Hadrian 17 = 132 A.D.

---

[1] For the exact title see the bibliography at the end s.v. Stobart [1].

[2] Brugsch [1] p. 19 writes "Je découvris dans cette collection quatre tablettes . . . représentées sur les planches II–V, annexées à ce mémoire" (Stobart [1]). At least my copy, however, of Stobart [1] does not contain any text beside the plates and all four tablets are reproduced together on Pl. II. Spiegelberg [3] p. 147 quotes also "Tafel 2" only.

[3] Brugsch [1] p. 20.
[4] Cf. Brugsch [1] p. 63. Hadrian's reign lasted 21 years.
[5] These results were announced in Neugebauer [1].
[6] Brugsch [1] p. 63 quotes a letter from Biot, informing him, that Mr. Ellis in London, assistant to Airy at Greenwich, calculated planetary positions from 105 A.D. to 114 A.D. and found them in agreement with the text. Those years correspond indeed to Trajan 9 to 18 and this is just the first part of tablets I and II. Satisfied by this coincidence further investigations have been omitted and therefore tablet III remained misplaced.

The "years" in the papyrus P.8279 run from "14" to "41" which excludes all emperors except Augustus, an assumption confirmed by the astronomical calculation (i.e. 16 B.C. to 11 A.D.).

2. The following two paragraphs (§ 2 and 3) contain the complete translation of the Berlin papyrus (in the following quoted as "P") and of the Stobart tablets (quoted as "S" or "S,A" to "S,E" respectively). The astronomical and historical commentary will be given in § 4 and remarks on writing and terminology in § 5.

Some general explanations with respect to restorations of destroyed passages, indicated by [ ], are necessary. There are three different kinds of restorations. First, trivial restorations: to this group belong the majority of restorations of missing months or zodiacal signs, because they are simply determined by their natural order. No consideration is paid to these cases in the apparatus criticus. There are however two more types, mentioned in the apparatus as "r.b.i." ("restoration by interpolation") and "r.b.c." ("restoration by calculation"), respectively. The r.b.i. are obtained by the following graphical method: the positions of the planets as given by the text were represented in a time-distance-diagram. Missing positions have then been found in this diagram by graphical interpolation by means of a curve congruent as much as possible to a part of a curve connecting points given by the text and in analogous relative position. It can be expected that restorations obtained by this method give within very narrow limits the same result as if they were obtained by using the original methods. This could be explicitly verified in different cases where numbers found in this way agreed perfectly with traces still visible on the texts.

In contrast to this are "r.b.c." obtained by *modern* calculation [7] which therefore cannot claim to reflect the original form of the texts. Hence the r.b.i. are favored as much as possible.

---

[7] For the calculation of the Julian dates Schram Tafeln and P. V. Neugebauer [2] have been used; for the calculation of the planetary-positions P. V. Neugebauer [1] and P. V. Neugebauer T. A. Chr.

§ 2. The Papyrus Berlin P.8279 rev.

P.8279 is at present about 25.5 cm high and $1\frac{3}{4}$ m long. The papyrus contains, in its present state of preservation, on its reverse 39 more or less damaged columns of ca 25 lines each. No margin is preserved at the beginning or at the end; furthermore one complete column is missing in a gap between columns "XXVIII" and "XXIX".[8]

The papyrus comes from the Fayūm and was written later than the second year of Claudius (42 A.D.) as is proved by the following fact: the obverse contains a Greek text [9] concerning delivery of grain, dated in the year mentioned; on the reverse there is only one small notice written in Greek,[10] belonging to the text on the obverse, and the Demotic text is written around the Greek signs.[11]

As far as is known to me P has been discussed only twice since its publication in 1902, namely by Oefele.[12] Both articles are mainly concerned with dates in the New Testament which have no more than a purely accidental connection with our text.

Translation

General remark: The original text does not contain any vertical or horizontal ruling or lines between the single groups referring to the same planet. For the meaning of "s.s." in the apparatus criticus see p. 245.

The sign □ following the year-numbers is a symbol for "Saturn," as Dr. G. R. Hughes discovered (see p. 247).

I thought it unnecessary to mention in the apparatus the numerous differences between the present and Spiegelberg's edition.

---

[8] In spite of this fact, I have retained Spiegelberg's numbering in the following in order to avoid confusion.

[9] Never published, as far as I know. Spiegelberg DPB plate 99 gives only a photograph of col. IV to VIII.

[10] In col. XXXV; cf. the apparatus criticus below p. 219.

[11] Spiegelberg DPB p. 29. In spite of his own remark about this fact, Spiegelberg gives on p. 29 and p. 36 the date "after Augustus year 41 = 10 A.D."

[12] Oefele [1] (1902) and [2] (1903).

Papyrus Berlin P.8279

Reverse	I			II			III			IV			V		
1.	[9]	16	♉	11	6	♌	[1]1	21	[♍]	6	30	♒	10	5	♋
	[♀ 1]	1	♎	year	15	□	☿ 1	11	♍	8	8	♓	12	18	♌
	[1]	26	♏	3	21	♐	1	30	♎	9	19	♈	year	17	□
	[2]	21	♐	♃ 6	8	♊	2	20	♏	11	4	♉	3	30	♑
5.	[3]	17	♑	7	27	♋	3	10	♐	12	30	♊	♃ 2	20	♍
	[4]	13	♒	♂ 5	9	♉	3	26	♏	♀ 1	12	♏	8	5	♌
	[5]	10	♓	7	5	♊	4	25	♐	.5	10	♐	9	29	♍
	[6]	9	♈	9	27	♋	5	25	♑	6	16	♑	♂ 8	7	♋
	[10]	7	♉	10	23	♌	6	11	♒	7	11	♒	10	9	♌
10.	[1]1	13	♊	12	15	♍	6	18	♓	8	6	♓	11	23	♍
	[1]2	8	♋	♀ [1]	24	♍	8	16	♈	9	1	♈	♀ 1	17	♎
	[epa]g.	5	♌	[2]	18	♎	9	8	♉	9	26	♉	2	11	♏
	[☿] 1	21	♍	[3]	13	♏	9	23	♊	10	21	♊	3	5	♐
	2	9	♎	4	7	♐	10	9	♋	11	15	♋	3	30	[♑]
15.	2	28	♏	4	29	♑	12	[13	♌]	12	7	♌	4	25	6
	3	17	♐	5	25	♒	year	16[□].		12	29	[♍]	[5]	20	♓
	5	26	♑	6	1[9]	[♓]	7	1	[♑]	☿ 1	8	♋	6	16	♈
	6	10	♒	7	15	[♈]	10	5	[♐]	2	16	♏	7	13	♉
	‘6	26	♓	8	10	[♉]	♃ 1	[15	♌]	4	24	♐	8	12	[♊]
20.	7	13	♈	9	5	[♊]	7	[5	♋]	[5]	9	♑	1[0]	17	[♋]
	9	17	♉	9	30	[♋]	8	[30	♌]	[5]	25	♒	1[1]	8	[♊]
	10	4	♊	10	[25	♌]	[♂] 1	[20	♎]	[6]	16	♓	1[2]	9	[♋]
	10	18	♋				[3	1	♏]	[8]	15	[♈]	☿ 1	20	[♍]
							[4	14	♐]	[9	2]	♉	1	[23	♎]
25.							[5	22	♑]	[9	19]	♊	[2	21	♍]

I,1 [9]:	refers to the last position of ♂ in the year (Augustus) 14. The following position (given in II,6) is the entrance of ♂ into ♉ coming from ♈. The sign ♉ in I,1 is therefore incorrect. Calculation gives for month IX day 16 the place ♓ 0.
II,8 9:	sic! instead of 8.
III,1 [♍]:	restoration practically certain because of visible remains of the sign. Thereafter a complete line is omitted, which may be restored as 12 19 ♎.
III,8 25:	sic! instead of 15.
III,9 11:	sic! probably instead of 1.
III,16 [□]:	traces visible; that 16 was followed by -t is highly improbable.
III,19 to 25:	dates r.b.i.
IV,17 ♋:	sic! instead of ♍. Next line missing, which may be restored as 1 27 ♎.
V,23 20:	sic! instead of 4.

Papyrus Berlin P.8279

Reverse	VI			VII			VIII			IX			X		
1.	2	21	♎	10	11	♉	7	13	♓	9	12	♈	7	2	♓
	3	29	♏	11	20	♊	8	3	♈	10	7	♉	8	11	♈
	4	15	♐	♀ 1	11	♌	8	18	♉	11	2	♊	9	22	♉
	5	2	♑	2	18	♍	9	11	♊	11	26	⊗	11	7	♊
5.	5	18	♒	3	2	[♎]	11	11	⊗	12	20	♌	12	26	⊗
	7	23	♓	3	26	♏	12	1	♌	☿ 2	24	♎	♀ 1	9	♍
	8	9	♈	4	1[9]	♐		2	♍	3	10	♏	2	3	♎
	8	24	♉	5	1[5	♄]	year	19	□	3	28	♐	2	27	♏
	9	14	♊	6	10	[♒]	6	6	♒	4	16	♄	3	21	♐
10.	11	22	⊗	7	4	[♓]	♃ 4	22	♏	6	25	♒	4	18	♄
	12	11	♌	7	25	[♈]	10	12	♎	7	9	♓	5	11	♒
	12	28	♍	8	21	♉	11	27	♏	7	25	♈	6	5	♓
	year	18	□	9	26	♊	♂ 1	26	⊗	8	11	♉	7	1	♈
	4	28	♄	10	11	⊗	9	3	♌	10	13	♊	7	2[6]	♉
15.	♃ 3	20	♎	11	2	♌	11	4	♍	11	5	⊗	8	20	[♊]
	1[1]	9	♍	11	2[9]	♍	12	20	♎	11	[20]	♌	9	20	[⊗]
	[♂ 1]	5	♎	12	25	♎	♀ 1	16	♏	12	10	♍	10	[22	♌]
	[2]	20	♏	☿ 1	9	♎	2	[13]	♐	year	20	□	11	1[9	⊗]
	[4]	2	♐	2	21	[♏]	3	12	♄	1	25	♒	☿ 2	1[3	♎]
20.	[5	10]	♄	3	6	♐	4	13	♒	♃ [4]	23	♐	3	3	[♏]
	[6	16	♒]	3	24	♄	[5]	30	♄	♂ 1	30	♏	3	22	[♐]
	[7	22	♓]	5	11	[♒]	[7]	6	♒	[3]	11	♐	4	1[4	♄]
	[8	30	♈]	[6	4	♄]	[8]	16	♓	4	12	♄	4	1[9	♐]
				[6	20	♒]				[5	22	♒]	6	[1	♄]
25.													[6	17	♒]

VI,1 2 21 : probably erroneous repetition of the numbers in the foregoing line (V,25). Instead of the above, 3 11 should be expected (the sign ♎ being correct).

VI,14/15 : one line missing, which indicates the entrance into ♐.

VI,15 20 : perhaps units after 20 to be restored, but no traces left.

VI,16 1[1] : only vague traces of 10 left.

VI,16 : one line missing, as the disorder in the zodiacal signs shows. A possible restoration would be

<9 9> ♍ (calculation: ♍ 28)
1[1] 9 <♎> (calculation: ♍ 29).

VI,20 to 23 : dates r.b.i.

VII,4 18 : s.s. for 8. The date 18 is incorrect (probably instead of 8).

VII,13 26 : sic! instead of 16.

VII,18/19 : one line omitted, probably 2 5 ♍.

VII,19 to 21 : according to the line omitted between line 18 and 19 all signs should be reduced by one as follows
2 21 [♎] (calculation: ♍ 28)
3 6 ♏ (calculation: ♎ 19)
3 24 ♐ (calculation: ♏ 20).

VII,21/22 : one line omitted, probably 4 16 ♄.

VII,22 11 : reading of the 10 almost sure, of the following 1 probable. Calculation gives ♒ 5 instead of ♒ 0.

VIII,5 (day) 11 : reading of the 1. after 10 doubtful.

VIII,7 2 : the text here omits the sign for "epag.", which omission happens nowhere else.

VIII,18 [13] : traces, compatible with 3, visible.

IX,15 5 : before 5 lacuna; 5 is written near to ⊗ and a little stroke is still visible, so that the reading [1]5 would be quite possible, but the interpolation obviously requires 5 and not 15.

IX,16 [20] : a restoration [22] or [23] would fit much better but the end of 20 is still visible and there is no trace of any unit.

IX,18 year 20 : ḥ3.t-sp 20-t.

X,10 18 : s.s. for 8.

X,15,16 20 : units could be restored but r.b.i. requires 20 only.

X,17,18 : r.b.i.

X,19 1[3 : traces of 3 visible.

X,24 [1 : traces visible.

Papyrus Berlin P.8279

Reverse

	XI			XII			XIII			XIV			XV		
1.	7	2	♓	12	14	♌	♀ 1	3	♎	11	12	♈	6	11	♓
	7	12	♈	☿ 1	20	♍	1	28	♏	♂ 1	24	♌	7	5	♒
	9	14	♉	2	7	♎	2	24	♐	3	11	♍	7	21	[♓]
	10	11	♊	2	26	♏	3	19	♑	10	25	♎	8	15	[♈]
5.	10	22	⊗	3	25	♐	4	15	♒	12	19	♏	8	30	♉
	11	9	♌	5	24	♑	5	12	♓	♀ 1	25	♍	9	16	♊
	year	21 □		6	9	♒	6	12	♈	2	15	♎	10	21	⊗
	8	14	♓	6	25	♓	9	1	♉	3	13	♏	12	16	♌
	♃ 5	18	♑	7	22	♈	11	15	♊	4	8	♐	epag.	4	♍
10.	♂ 2	18	12	9	17	♉	12	11	⊗	5	1	♑	year	24 □	
	10	12	♍	10	1	♊	epag.	5	♌	5	25	♒	7	4	♈
	[1]2	3	♎	10	15	⊗	☿ 1	13	♍	6	19	♓	♃ 1	8	♓
	♀ [1]	18	♌	11	16	♌	2	2	♎	7	15	♈	6	5	♈
	2	19	♍	year	22 □		2	21	♏	8	11	♉	11	5	♉
15.	[2]	14	♎	5	2[7]	♓	4	29	[♐]	9	6	♊	♂ 1	26	♐
	4	18	♏	♃ 6	3	♒	5	11	[♑]	10	1	⊗	3	3	♑
	5	9	♐	♂ 1	13	♏	6	2	[♒]	10	27	♌	4	10	[♒]
	6	1	♑	2	2[4]	♐	6	17	♓	11	23	♍	5	20	♓
	6	29	♒	4	3	♑	8	20	♈	12	27	♎	7	5	♈
20.	7	22	♓	5	8	♒	9	7	♉	☿ 1	7	♍	8	15	♉
	8	16	♈	6	14	♓	9	2[5]	♊	1	26	♎	9	30	♊
	9	11	♉	7	[22]	♈	1[0]	18]	⊗	2	14	♏	11	15	⊗
	10	[27]	♊	9	[6	♉]	1[2]	2[1]	♌	4	22	♐	epag.	4	[♌]
	1[1]	20	[⊗]	10	1[8	♊]	[year]	23 □		5	6	♑	♀ 1	15	♐
25.				[12	8	⊗]	[12]	5	[♓]	[5]	26	♒	5	13	[♐]
							[♃ 6]	8	[♓]						

XI,2 12: sic! about 18 should be expected.

XI,4 11: sic! about 3 should be expected.

XI,8: one line omitted, after line 8, perhaps <12 15 ♒>. Calculation gives ♒ 24.

XI,15 [2]: this restoration of the traces almost certain, in spite of the fact that [3] would be correct (the 2 could be explained as repetition of the preceding 2).

XI,23 [27]: traces visible.

XI,23 ♊: sic! instead of ⊗. There is one foregoing line omitted, perhaps 10 5 ♊.

XI,24 [⊗]: doubtless correct restoration in view of the foregoing ♊ and the following ♌ in spite of the fact that [♌] would be correct (see note to XI,23).

XII,1 ♌: sic! instead of ♍; see note to XI,23 and 24.

XII,15 2[7]: only 2[6] or 2[7] compatible with traces.

XII,22 [22]: r.b.i.

XIII,25 [♓]: traces visible.

XIII,26 6]: r.b.c. gives ♒ 27; traces would fit better to month 4 or 8.

XV,24 ♐: sic! instead of ♎.

XV,25 13: 3 clear, 10 damaged but better than 20. Calculation gives for month 5 day 13 the position ♏ 26.

XV,25 [♐]: this restoration fits better to the preserved traces than ♑, which should be expected if the mistake from the foregoing line had been continued.

Papyrus Berlin P.8279

Reverse	XVI			XVII			XVIII			XIX			XX		
1.	[6]	18	♑	♂ 2	[15	♍]	[8	15	♉]	[8	1	♈]	[♂ 1]	15	♍
	7	18	♒	4	5	♎	10	10	[♊]	8	[26	♉]	[2]	14	♎
	8	8	♓	6	9	♏	11	14	[⊗]	9	[20	♊]	[3]	29	♏
	9	3	♈	9	29	♎	11	28	♌	10	[14	⊗]	[6]	3	♐
5.	9	28	♉	10	28	♏	12	14	♍	11	[8	♌]	[12]	10	♑
	10	22	♊	epag.	5	♐	year	26	□	12	2	[♍]	♀ 1	18	♏
	11	16	⊗	♀ 1	20	♎	8	22	♉	12	27	[♎]	[2]	15	♐
	12	9	♌	2	23	♏	♃ 2	29	♉	♄ 2	21	[♎]	[3]	19	♑
	epag.	2	♍	3	7	♐	6	7	♊	3	8	[♏]	4	18	♒
10.	☿ 1	18	♎	4	2	♑	12	1	⊗	3	28	♐	[5	2]6	♑
	3	28	♏	4	27	♒	♂ 2	10	♑	4	15	♑	[6]	9	♒
	4	12	♐	5	22	♓	3	15	♒	5	11	♐	[8]	18	♓
	4	[2]9	[♑]	6	18	♈	4	23	♓	6	3	♑	[9]	11	♈
	5	[1]6	♒	7	16	♉	6	4	♈	6	25	♒	[10	7]	♉
15.	6	22	♓	8	15	♊	7	6	♉	7	23	♓	[11]	3	♊
	8	7	♈	9	23	⊗	9	6	♊	7	29	♈	[11]	28	⊗
	9	10	♊	10	9	♊	10	22	⊗	8	13	♉	[12]	22	♌
	[10]	20	⊗	12	12	⊗	12	13	♌	10	18	♊	☿ [1]	2	♌
	[12	5]	♌	☿ 1	9	♎	♀ 1	13	♌	11	3	⊗	[1]	24	♍
20.	[1]2	25	♍	2	8	♍	2	9	♍	11	16	♌	2	13	♎
	year	[2]5	□	2	28	♎	3	4	♎	12	13	♍	3	2	♏
	2	8	♈	3	18	♏	3	28	♏	year	27	□	3	21	♐
	♃ 4	1	♈	4	4	♐	4	23	♐	1	15	♉	5	11	♑
	6	3	♉	4	22	♑	5	18	♑	♃ 12	29	♌	[6]	1[5]	♒
25.	[11	1]8	♊	6	7	♒	6	12	♒						
				7	12	♓	[7	6]	♓						
				[7	29]	♈									

XVI,11 28 :	s.s. for 8.
XVI,16/17 :	one line omitted, perhaps 8 23 ♉.
XVI,22 8 :	traces only.
XVI,25 1]8 :	s.s. for 8.
XVII,8 23 :	sic! instead of 13.
XVII,16 23 :	sic! instead of 13.
XVII,22 18 :	s.s. for 8.
XVII,27 29] :	r.b.i.
XVIII,1 15 :	r.b.i.
XVIII,15 6 :	sic! instead of about 20.
XVIII,24 18 :	s.s. for 8.
XIX,1 [8 :	traces.
XIX,1 to 5 :	r.b.i.
XIX,10 28 :	s.s. for 8.
XIX,15 23 :	sic! instead of 13.
XIX,25 :	as far as the damaged margin enables us to judge, this line was empty. This makes improbable the assumption that a retrograde entrance into ⊗ was given here, which could explain the repetition of the sign ♌ in XXI,9 (see note there).
XX,2 14 :	sic! instead of 24.
XX,4 3 :	only traces preserved, which could also be restored as 1.
XX,9 18 :	s.s. for 8.
XX,16 28 :	s.s. for 8.

Papyrus Berlin P.8279

Reverse	XXI			XXII			XXIII			XXIV			XXV		
1.	[7]	1	♓	8	24	♊	4	1[1	♏]	[6	1]7	♓	☿ 1	6	♍
	7	17	♈	9	20	⊗	5	26	♐	8	[20]	♈	[1]	25	≏
	9	21	♉	10	18	Ω	7	8	♑	9	5	♉	[2]	14	♏
	10	8	♊	11	28	♍	8	15	♒	9	20	♊	2	29	≏
5.	10	[2]1	⊗	♀ 1	19	♍	9	28	♓	10	10	⊗	4	3	♏
	11	10	Ω	2	7	≏	12	4	♈	1[2]	27	Ω	4	19	♐
	year	2[8]	□	2	26	♏	♀ 1	13	Ω	year	30	□	5	6	♑
	10	[3]	♊	3	14	♐	2	12	♍	[1]1	26	⊗	6	28	♓
	♃ 4	[3]	Ω	4	4	♏	3	16	≏	♃ [3]	4	≏	8	13	♈
10.	♂ 2	10	♒	4	30	♐	4	14	♏	♂ [2]	10	♓	8	29	♉
	3	[1]8	♓	5	23	♑	5	10	♐	[3]	24	♈	9	14	♊
	5	10	♈	6	8	♒	6	4	♑	[5]	30	♉	11	25	⊗
	6	25	♉	6	24	♓	6	29	♒	7	19	♊	12	25	Ω
	8	14	♊	7	15	♈	7	24	♓	9	10	[⊗]	epag.	2	♍
15.	10	[6	⊗]	9	1[3]	♉	8	18	♈	11	4	[Ω]	year	31	□
	11	[29	Ω]	9	28	♊	9	14	♉	12	2[2	♍]	6	18	♊
	♀ 1	12	[♍]	10	[1]4	⊗	10	8	♊	♀ 1	[5	≏]	7	12	⊗
	2	5	≏	11	16	Ω	10	29	⊗	1	29	♏	♃ 3	9	♏
	2	29	♏	12	1	⊗	11	19	Ω	2	24	♐	♂ 2	5	≏
20.	3	25	♐	12	28	Ω	12	15	♍	3	20	♑	3	18	♏
	4	21	[♑]	year	29	□	♀ 1	14	♍	4	16	♒	4	30	♐
	5	13	♒	2	13	♊	2	2	≏	5	14	♓	6	9	♑
	6	8	♓	♃ 1	29	♍	2	21	♏	6	13	♈	7	15	♒
	7	[4]	♈	♂ 1	10	♍	4	29	♐	10	14	♉	8	24	♓
25.	7	[29	♉]	1	26	≏	5	14	♑	11	18	♊	10	6	♈
							5	30	♒	[12]	12	⊗	[11	2]6	♉

XXI,7 2[8]: traces of 8 visible, certainly not followed by -t.
XXI,8 and 9 [3]: very little visible.
XXI,9 Ω: in col. XIX,24 the entrance into Ω is already correctly indicated (calculation gives ⊗ 28) so that here either ⊗ or ♍ should be expected. The sign ♍ can be excluded because entrance into ♍ first occurs in the next year (see XXII,23 in accordance with calculation which gives Ω 27). Therefore retrograde entrance into ⊗ could only be meant, but calculation gives for month 4 day 3 the place Ω 11 (still in direct movement); ♃ reaches the stationary point 6 days later but does reach Ω 0 in its following retrograde movement.

XXI,24 [4]: traces visible.

XXII,15 1[3]: r.b.i.
XXII,17 [1]4: instead of 4 also 3 possible, but no other sign.

XXII,21 year 29: ḥ3.t-sp 29-t.
XXII,25 1: sic! instead of 2.
XXIII,21 ☿: the sign stands on the level between lines 21 and 22.

XXIV,2 [20]: space for one sign only; restoration therefore certain.

XXIV,7 year 30: below the sign 30 a blot, which could be interpreted as -t (cf. IX,18 and XII,21).

XXIV,8 [1]1: no traces of 10 visible but required by calculation.

XXIV,17 [5: traces visible, which only conform to 1, 2 or 5 and 5 is best by interpolation.

XXIV,25 18: s.s. for 8.
XXV,7/8: one line omitted, perhaps 5 22 ♒.
XXV,8 6: sic! instead of 7 as calculation shows (month 6 gives ♒ 9 but month 7 correctly ♓ 1).

XXV,8 28: s.s. for 8.
XXV,12 25: sic! instead of 15.

Papyrus Berlin P.8279

Reverse

Line	XXVI				XXVII				XXVIII			XXIX		
1.	♀	[1	1]	♌	11	21	⊕		6	29	[≈]	3	6	♎
		1	2[5]	♍	12	7	♌		7	14	[♓]	4	1	♏
		2	21	♎	year	32	□		7	30	♈	4	25	♐
		3	16	♏	3	24	⊕		8	20	♉	5	20	♑
5.		4	10	♐	♃ 4	18	♐		10	23	♊	6	15	≈
		5	3	♑	♂ 6	18	♊		11	11	⊕	7	6	♓
		5	28	≈	8	20	⊕		11	23	♌	8	[1]	♈
		6	21	♓	10	14	12		12	13	♍	8	[26]	♉
		7	19	♈	12	5	♍		year	33	□	[9	20]	♊
10.		[8]	13	♉	♀ 1	18	♏		1	26	[♌]	[10	13]	⊕
		9	8	♊	5	18	4		[♃] 5	[////	♑]	[11	6]	♌
		10	3	⊕	6	19	5		////	////	[♐]	[11	30]	♍
		10	29	[♌]	7	15	≈		////	////	[♑]	[12	28]	♎
		11	21	[♍]	8	10	♓		[♂]////	////	////	[♉ ////	////]	♍
15.		12	23	[♎]	9	5	♈		////////////////////////////			////	////	♎
	☿	1	13	[♎]	9	29	♉					////	////	♏
		3	25	[♏]	10	24	♊					////	////	♐
		4	10	[♐]	11	18	⊕		destroyed			////////////////////////////		
		4	28	[♑]	12	11	♌							
20.		5	11	≈	epag.	5	♍							
		6	10	♑	☿ 1	14	♎					destroyed		
		7	3	≈	1	24	♍							
		7	21	♓	2	28	♎							
		8	6	♈	3	24	♏							
25.		8	21	♉	4	2	[♐]							
		9	22	[♊]	[4	18	♑]							

(between XXVIII and XXIX: *one column destroyed*)

XXVI,7 28 :   s.s. for 8.

XXVII,2/3 :   one line missing, perhaps 12 23 ♍.
XXVII,5,6,10,11 18:   s.s. for 8.
XXVII,24 24 :   sic! instead of 14.
XXVII,26 [4 :   traces visible.

XXVIII,10 [♌] :   restoration certain, because clear traces visible.

XXVIII,11 [♃] :   no traces of the sign ♃ visible although only half of the necessary space on the papyrus missing.

XXVIII,14 [♂] :   restoration of traces certain. The space between ♂ and the 5 in line 11 is very probably two lines, certainly not less.

XXIX,7 [1] :   traces visible.
XXIX,8 to 13 :   r.b.i.

Papyrus Berlin P.8279

Reverse

Line	XXX	XXXI	XXXII	XXXIII	XXXIV
1.	11 16 ♌		[3] 23 ♐	♂ 1 24 ♏	[5] 3 [♑]
2.	year 35 □		[4] 21 ♑	4 9 ♐	5 21 [♒]
3.	1 9 ♌		[5] 15 ♒	4 13 ♑	7 28 ♓
4.	♃ 6 4 ♓	destroyed	[6] 9 ♓	5 16 ♒	8 12 ♈
5.	10 12 ♈		[7] 3 [♈]	6 26 ♓	8 [29] ♉
6.	♂ 2 11 ♏		[7] 29 ♉	8 6 ♈	9 15 ♊
7.	3 24 ♐	////////////////////////////	[8] 26 [♊]	9 16 ♉	11 30 [⊗]
8.	5 3 ♑	//// //// [♈]	[9] 22 ⊗	10 28 ♊	12 14 ♌
9.	6 7 ♒	//// //// [♉]	10 21 ♌	12 16 ⊗	12 30 ♍
10.	7 12 ♓	//// //// [♊]	11 30 ♍	♀ 1 10 ♌	year 38 □
11.	8 24 ♈	//// //// [⊗]	☿ 1 12 ♍	2 8 ♍	3 7 ≏
12.	10 9 ♉	//// //[4 ♌]	2 1 ≏	3 [18 ≏]	8 18 [♍]
13.	11 28 ♊	[year 36 □]	2 20 ♏	[4 13 ♏]	11 27 ≏
14.	♀ 1 20 ♏	[1 ////] ♍	3 28 ♐	[5 6 ♐]	♃ 1[1] 14 ⊗
15.	2 18 ♐	[♃ 1 2]1 ♓	5 12 ♑	[5 30 ♑]	♂ 2 5 ♌
16.	3 16 ♑	[6 5] ♈	5 28 ♒	[6 24 ♒]	4 5 ♍
17.	[4] 21 ♒	[10 25] ♉	6 16 ♓	[7 18 ♓]	11 20 ≏
18.	[6 2] ♑	[♂ 1 30 ⊗]	8 18 ♈	[8 13 ♈]	epag. 5 ♏
19.	[7 10] ♒	[4 26] ♌	9 [3] ♉	[9 7 ♉]	♀ 1 6 ≏
20.	[8 13] ♓	[6] 10 ⊗	[9] 20 ♊	10 3 ♊	2 1 ♏
21.	/////////////////////////	8 2[7] ♌	[10 1]3 ⊗	[1]1 1 ⊗	2 26 ♐
22.		10 24 ♍	1[2] 21 ♌	☿ 1 20 ♍	3 22 ♑
23.	destroyed	1[2] 16 ≏	year 37 □	[1 2]2 ≏	4 18 ♒
24.		♀ 1 [4 ♍]	1 1 ♍	[2 10] ♏	5 13 ♓
25.		2 [2 ≏]	[♃ 10 2]4 ♊	[4 18] ♐	6 [14 ♈]
		3 [1 ♏]			

XXX,13 28: s.s. for 8.
XXX,15 18: s.s. for 8.

XXXI,8 to 12: traces visible.
XXXI,15 to 17: r.b.c. gives: 1 21 ♓ 26 retrograd
                      6 5 ♓ 28 direct
                    10 25 ♈ 27 direct
XXXI,20 [6]: traces visible.

XXXII,11/12 ☿: the sign stands on the level between line 11 and 12.
XXXII,19 [3]: traces visible.
XXXII,25 10 2]4: r.b.c. which gives ♉ 25; traces of 4 visible.

XXXIII,2 4: sic! instead of 3.
XXXIII,8 28: s.s. for 8.

XXXIII,21: two more lines for ♀ omitted, perhaps to be restored as
             11 23 ♌
             12 15 ♍
XXXIV,3 28: s.s. for 8.
XXXIV,5 [29]: traces of 9 visible.
XXXIV,7 30: this seems to be the best interpretation of the traces; 10 can not be excluded with certitude. Calculation gives for month 11 day 30 the place ⊗ 3 and also very good agreement for the preceding and following dates (♊ 1 and ♌ 1, respectively).
XXXIV,10 year 38: ḥ3.t-sp 38-t.
XXXIV,13 27: traces of 7 only.
XXXIV,14 1[1]: r.b.c. (♊ 25).
XXXIV,15 2: only one stroke preserved but reading 2 necessary.

Papyrus Berlin P.8279

Reverse	XXXV			XXXVI			XXXVII			XXXVIII			XXXIX		
1.	♏[10	12	♉]	[7	1]8	[8]	[11	7	11]	3	[17]	4	4	[30	6]
	11	18	♊	[8]	28	9	11	21	[12]	6	[1]	5	5	[25	7]
	1[2	1]3	⊗	10	13	10	12	10	[1]	6	[20]	6	6	[22	8]
	☿ 1	13	♎	♀ 1	3	12	year	40 □		7	6	7	7	19	[9]
5.	3	22	♏	1	28	♍	2	5	[2]	7	24	8	8	14	[10]
	4	9	♐	2	2[3]	♎	♃ 4	16	[12]	8	26	9	12	18	[11]
	4	2[8]	♑	3	18	♏	♂ 1	11	[12]	10	14	10	☿ 1	24	[1]
	5	18	♒	4	[13]	4	2	2	1	10	26	11	2	1[2	2]
	5	28	♑	5	20	5	4	28	2	11	14	12	2	[30	3]
10.	7	4	♒	5	28	6	12	1	3	[year]	41 □		3	1[9	4]
	7	19	♓	6	26	7	♀ 1	20	3	2	30	3	4	1[9	3]
	8	5	♈	7	[21]	8	5	19	4	[♃] 1	12	[1]	[5]	4	[4]
	8	29	♉	[9	11]	10	6	21	5	♂ 1	13	[4]	/////////////////////////		
	10	18	♊	[10	7	11]	7	15	6	2	[20	5]			
15.	11	19	⊗	[11	3	12]	8	11	7	3	[30	6]	destroyed		
	12	3	♌	[11	28	1]	9	6	8	5	[12	7]			
	12	19	♍	[12	23	2]	9	30	9	6	[22	8]			
	year	39 □		[☿ 2	27	2]	10	25	10	8	[4	9]			
	1	23	♎	[3	15]	3	11	9	11	9	[16	10]			
20.	♃ 12	12	12	[4	3]	4	12	12	12	[10	28	11]			
	♂ 2	12	4	[4]	20	5	☿ 1	11	12	[12	10	12]			
	3	21	5	6	28	6	1	26	1	♀ [1	1	1]			
	[4]	28	6	7	12	7	2	[14]	2	[1	25	2]			
	[6	8	7]	7	28	8	3	[1	3]	[2	19	3]			
25.				8	22	9				[3	13	4]			
				[10]	23	10				[4	7	5]			

XXXV,3 1]3 : reading 3 of traces plausible. Calculation gives for day 13 the position ♓ 23, and therefore for day 23 practically exactly entrance into ⊗; but both the foregoing and following positions give 6 or 7 degrees too much, so that day 13 is more probable than day 23.

XXXV, 6/7: between these two lines is to be found the Greek notice which is mentioned above p. 211 and which Schubart (Spiegelberg DPB p. 29 note 2) reads as "κυά(μου) followed by numbers." I confess that the only letter I am able to read is ν as the first letter.

XXXV,8 18 : s.s. for 8.
XXXV,9 28 : s.s. for 8.
XXXV,14 18 : s.s. for 8.
XXXV,18 year 39 : ḥ3.t-sp 39-t.
XXXV,23 28 : s.s. for 8.

XXXVI,3/4 : one line omitted, perhaps 11 29 11.
XXXVI,7 18 : s.s. for 8.

XXXVI,9 20 : sic! probably instead of 10.
XXXVI,12 [21] : traces visible.
XXXVI,12/13 : one line omitted, perhaps 8 16 9.
XXXVI,14 11] : traces, difficult to bring into agreement with 11 or ⊗; perhaps only blots.

XXXVII,4 40 : practically certain without -t.
XXXVII,6 [12] : restoration of this sign (= ♌) in analogy to year 28 (XXI,9); cf. commentary p. 233.
XXXVII,8 2 2 : sic! instead of 3 2.
XXXVII,19 9 : sic! instead of 19.

XXXVIII,6 8 : sic! instead of 9.
XXXVIII,12 [1] : traces visible.
XXXVIII,14 to 21 : the restoration here is an extrapolation checked by calculation and agreeing with the numbers of the months preserved.
XXXVIII,22 to 26 : r.b.i.

XXXIX,12 4 : reading very probable because of traces preserved.

§ 3. The Stobart-Tablets

The Stobart-tablets were purchased at Thebes in 1854/85.[13] The construction of the tablets is as follows: a wooden frame of about 8 by 12 *cm* supports a thin wooden foil covered with white plaster on both sides.[14] One side of the frame contains three pairs of holes to permit a number of tablets to be bound together into a kind of "book." The "pages" have a little rectangular thickening in the middle, also covered with plaster,[15] probably intended to prevent the wooden foils from warping.

Each side of a tablet contains five columns, separated from each other by fine double-lines in red. Horizontal lines enclose the groups referring to a single planet in each year, the "years" being written in red. Every column contains about 30 lines (not ruled), except the obverse of tablet A, where the lower part of this page is separated by a horizontal wooden fillet and the remaining small rectangular field was covered with plaster, which is now almost completely gone. It is therefore impossible to tell why this space was separated from the main text in the upper part of the tablet, from which nothing is missing. One could assume that tablet A, now the first of the preserved tablets, was actually the first page in a book and consequently gave some kind of "title" on its obverse.

---

[13] Stobart [1] plate II and Brugsch, Thes. I p. 64. Cf. furthermore Griffith [2] p. 71 note 2.

[14] This corresponds to the Roman "tabula cerussata" (cf. RE 3, 927).

[15] The script does not pay attention to these thickenings, except on tablet E reverse, where this space was left empty.

This is, however, very unlikely in view of the fact that tablet A does not begin with a new year, but somewhere at the end of "year 3" (of Vespasian).

All the preserved tablets were written by the same scribe as the ductus clearly shows.[16] The actual state of preservation of the tablets is fairly good. Tablet A is damaged in obverse col. V and reverse col. I/II by the scaling off of the plaster.[17] The obverse of tablet E got wet and most of the signs are therefore only visible in traces.[18] Compared with P however, the Stobart-tablets are not only much better preserved but also much more carefully written.

After the first excellent discussion of the Stobart-tablets by Brugsch in his "Nouvelles recherches . . ." (1856) no one attempted an investigation of these texts and one would try in vain to find them even mentioned in modern books on the history of science.[19]

Translation

Numbers printed in *italics* represent the special signs for dates; cf. moreover p. 244.

---

[16] There are however some small differences in the script which are sufficient to show that some time elapsed from tablet to tablet (cf. p. 247).

[17] One little fragment of plaster is now pasted on erroneously and upside down (also in Brugsch's publication) Here, of course, it is reproduced at the right place (cf. p. 222 note to II,4).

[18] Disregarded in Brugsch [1].

[19] Brugsch, of course, quoted these texts in his Thes. I (p. 64) and Aeg. (p. 325); they are not mentioned in Spiegelberg's edition of P but appear in Spiegelberg [1]. My attention was called to them by Dr. A. Volten in Copenhagen.

Stobart A

Obverse

	I			II			III			IV			V		
1.		☿		7	19	♈	8	15	♈	9	2	♈		♃	
	1 ☉	13	♍	8	27	♉	9	1	♉	10 ☉	4	♉	1 ☉	1	♏
2.	2	2	♎	11	11	♊	☉	17	♊	11	9	♊	4	21	♐
	☉[2]	1	♏	11	27	♋	10 ☉	5	♋	12	4	♋		♂	
5.	3	11	♐		♀		[1]2	7	♌	☉	28	♌	1 ☉	25	♐
	5	17	♑	1 ☉	1	♌		year	5		☿		3	6	♑
	6	4	♒	2	11	♍		b		1 ☉	1	♍	4	14	♒
	☉	20	♓	3	9	♎	1 ☉	1	♏	☉	20	♎	5	22	♓
	7	9	♈	4	4	♏	5	5	♐	2 ☉	7	♏	7	2	♈
10.	8	11	♉	☉	29	♐	9	25	♏	3	8	♎	8	10	♉
	9	15	♊	5	23	♑		♃		☉	13	♏	9	24	♊
	10	11	♋	6	18	♒	1 ☉	1	♎	4	14	♐	11	5	♋
	11	4	♌	7	12	♓	4	6	♏	5 ☉	1	♑	12	29	♌
	12	12	♋	8	7	♈	9	16	♎	☉	18	♒		♀	
15.	☉	12	♌	9	1	♉	12	16	♏	6	8	♓	1 ☉	18	♍
		year	4	☉	26	♊		[♂]		☉	27	♒	2	12	♎
		b		10	20	♋	1 ☉	16	♌	7	17	♓	[3	7	♏]
	[1] ☉	1	♏	11	14	♌	3	11	♍	8 ☉	8	♈	[4	1	♐]
		♃		12	8	♍	5	6	♎	☉	22	♉	[☉	26	♑]
20.	[1] ☉	1	♍	epag. ☉	3	♎	☉	22	♍	9 ☉	10	♊	[5	20	♒]
	[1]1	24	♎		☿		11	1	♎	10 ☉	5	[♋]	[6	15	♓]
		♂		1 ☉	1	♌	12	19	♏	12	10	♌	[7	9	♈]
	[1] ☉	7	♏	☉	7	♍		♀		☉	28	♍	[8	3	♉]
	[2]	18	♐	☉	26	♎	1 ☉	24	♏		year	6	[☉	27	♊]
25.	[3]	29	♑	2	15	♏	2	17	♐		b		[9	22	♋]
	[5]	6	♒	4	20	♐	3	12	♑	1 ☉	1	[♏]	[10	17	♌]
	[6]	10	♓	5	8	♑	4	7	♒	2	7	♐	1[1	12	♍]
				☉	23	♒	5	11	♓				12	13	[♎]
				6	12	♓									

A obverse.

I,4   ☉ [2]1 :   Brugsch gives "détruit" as month and "11" as day; but ☉ is certain and the following traces are doubtless remains of 20, as calculation demands.

I,10,11 :   calculation gives

    8   11   )(24   (about stationary)
    9   15   ♉ 2   (direct)

Apparently two lines are omitted and the text may be restored as

    7   9   ♈
    8   11   <)(>
   <8   15   ♈>
    9   15   <♉>
   <9   30>   ♊
   10   11   ⊗

I,15   12 :   sic! instead of 22. The mistake is apparently caused by the preceding 12. Calculation shows that 22 fits very well with entrance into ♌, as well as the preceding 12 as the date for the retrograde entrance into ⊗. The correction 22 instead of 12 has already been proposed by Brugsch.

I,18   [1] :   traces visible.

I,21   [1]1 :   1 clear and little trace of 10 visible.

II,3   11 :   sic! instead of (month) 10.

III,5   7 :   The scribe first wrote the ordinary number sign for 7 and intended thereafter to correct it to the date-number sign but wrote the first part only, probably in order to avoid further illegibility.

III,18   3 :   Brugsch erroneously 4.

III,22   19 :   Brugsch erroneously 9.

III,25   17 :   Brugsch erroneously 12.

III,27   7 :   Brugsch erroneously 2.

V,21   )(] :   traces visible.

V,24   ♊] :   traces visible.

V,25   ⊗] :   traces visible.

V,28   ≏] :   traces visible.

Stobart A

Reverse

	I	II	III	IV	V
1.	[♉]	[7 3 ♒]	4 26 ♓	year 9	☉ 12 ♍
	[1 ☉ 1 ♍]	[☉ 28 ♓]	6 7 ♈	♄	2 ☉ 6 ♎
	[☉ 13 ♎]	[8 22 ♈]	7 18 ♉	1 ☉ 1 ♑	☉ 25 ♏
	[2 3 ♏]	[9 18] ♉	9 4 ♊	♃	3 15 ♐
5.	[4 8 ♐]	[10 1]2 ♊	[10] 17 ♋	1 ☉ 1 ♒	5 19 ♑
	[☉ 30 ♑]	[11] 6 ♋	[1]2 5 ♌	6 14 ♓	6 7 ♒
	[5 21 ♒]	[☉ 30 ♌]	♀	♂	☉ 22 ♓
	[7 19 ♓]	[12 24 ♍]	1 ☉ 14 ♎	1 ☉ 23 ♍	7 11 ♈
	[8 9 ♈]	[♉]	2 6 ♏	3 11 ♎	9 15 ♉
10.	[☉ 28 ♉]	[1 ☉ 6] ♎	3 1 ♐	4 29 ♏	☉ 29 ♊
	[9 16 ♊]	[☉ 26] ♍	☉ 25 ♑	6 12 ♐	10 15 ♋
	[11 11 ♋]	[2 15] ♎	4 20 ♒	8 13 ♑	11 3 ♌
	[☉ 28 ♌]	[3 3] ♏	5 15 ♓	♀	year 10
	[12 15 ♍]	[☉ 21 ♐]	6 10 ♈	1 ☉ 6 ♌	♄
15.	[year 7]	[4 9 ♑]	7 8 ♉	2 ☉ 1 ♍	1 ☉ 1 ♑
	[♄]	[6 11 ♒]	[8 8] ♊	☉ 26 ♎	7 1 ♒
	[1 ☉ 1 ♐]	[☉ 26 ♓]	[10 ☉ 2] ♉	3 20 ♏	12 [3 ♑]
	[♃]	[7 14 ♈]	11 9 ♊	4 15 ♐	♃
	1 ☉ 1 [♐]	[8 4 ♉]	12 16 ♋	5 9 ♑	1 ☉ 1 [♓]
20.	5 12 ♑	[10 ☉ 4 ♊]	☿	6 3 ♒	6 23 [♈]
	♂	[☉ 22 ♋]	1 ☉ 1 ♍	☉ 27 ♓	11 1[1 ♉]
	1 ☉ 1 ♌	[11 10 ♌]	2 12 ♎	7 21 ♈	♂
	2 18 ♍	[12 1 ♍]	☉ 23 ♏	8 15 ♉	1 ☉ 1 [♑]
	4 6 ♎	[year 8]	3 16 ♐	9 9 ♊	2 1[7 ♒]
25.	6 14 ♏	[♄]	4 8 ♑	10 ☉ 2 ♋	3 2[7 ♓]
	epag. ☉ 3 ♐	1 [☉ 1 ♐]	5 4 ♐	☉ 28 ♌	5 7 [♈]
	♀	4 [18 ♑]	☉ 17 ♑	11 23 ♍	6 2[3 ♉]
	1 ☉ 1 ♎	[♃]	8 7 ♉	12 18 ♎	8 8 [♊]
	4 7 ♏	[1 ☉ 1 ♑]	10 11 ♊	☿	9 20 [♋]
30.	5 13 ♐	5 [☉ 30 ♒]	☉ 25 ♋	1 ☉ 1 ♌	11 1[0 ♌]
	6 9 ♑	♂	11 10 ♌		♀
		1 ☉ 1 ♐	12 9 ♍		
		2 10 ♑	☉ 29 ♌		
		3 18 ♒			

A reverse.

I,1 to 14 : r.b.i. only in order to take account of a possible filling of the available space.

II,1 ♒] : end of one horizontal line visible (disregarded by Brugsch) which may belong to the sign ♒, although traces of the other lines should also be expected on the preserved piece of plaster.

II,4 to 6 : the signs and numbers which are here preserved belong to a little piece of plaster which has been broken off and thereafter pasted in wrong position near to line 17 between cols. I and II.

II,10 to 23 : r.b.i. only.

II,17 ♓] : traces visible.

II,23 : Brugsch assumes three lines less, but the restoration given here is doubtless correct, within a margin of a few days error in the dates.

III,16 f. : r.b.i.

III,23 23 : sic! instead of 28 or 29.

V,16 ♒ : only one line preserved.

V,17 [3 : reading of small traces very uncertain.

V,26 7 : Brugsch "10 +?", but the remains fit much better to 7 than to 10.

Stobart C₁

Obverse

Line	I	II	III	IV	V
1.	4 19 ♐	1 16 ♍	12 10 ♌	☿	♀
	5 9 ♑	2 10 ≎	*⊙ 28* ♍	*1 ⊙ 1* ♍	1 9 ≎
	25 ♒	3· 4 ♏	year 10	*⊙ 11* ≎	2 4 ♏
	·6 12 ♓	29 ♐	♄	3 17 ♏	*⊙ 28* ♐
5.	8 14 ♈	4 22 ♑	*1 ⊙ 1* ♑	4 4 ♐	3 23 ♑
	9 1 ♉	5 16 ♒	♃	*⊙ 23* ♑	4 18 ♒
	⊙ 17 ♊	6 10 ♓	*1 ⊙ 1* ♍	5 11 ♒	5 13· ♓
	10 ⊙ 6 ♋	7 4 ♈	♂	7· 14 ♓	6 8 ♈
	12 17 ♌	*⊙ 29* ♉	*1 ⊙ 1* ♌	8 1 ♈	7 11 ♉
10.	epag. ⊙ 5 ♍	8 24 ♊	2 11 ♍	*⊙ 17* ♉	11 14 ♊
	year 9	9 18 ♋	3 29 ≎	9 6 ♊	12 14 ♋
	♄	10 13 ♌	5 27 ♏	11 14 ♋	☿
	1 ⊙ 1 ♐	11 8 ♍	12 28 ♐	12 2 ♌	1 5 ≎
	3 ⊙ 5 ♑	12 16 ≎	♀	*⊙ 18* ♍	*⊙ 12* ♍
15.	♃	☿	*1 ⊙ 1* ≎	year 11	2 16 ≎
	1 ⊙ 1 ♌	*1 ⊙ 19* ≎	2 18 ♍	♄	3 8 ♏
	12 28 ♍	2 9 ♏	*⊙ 23* ≎	*1 ⊙ 1* ♑	27 ♐
	♂	3 3 ≎	4 11 ♏	5 14 ♒	4 15 ♑
	[1] 16 ♐	*⊙ 17* ♏	5 11 ♐	♃	6 20 ♒
20.	2 27 ♑	4· 13 ♐	6 6 ♑	1 22 ≎	7 3 ♓
	4 5 ♒	*⊙ 30* ♑	7 1 ♒	♂	*⊙ 23·* ♈
	5 13 ♓	5 18 ♒	*⊙ 25* ♓	*1 ⊙ 1* ♐	9 7 ♉
	6 23 ♈	6 9 ♓	8 19 ♈	2 5 ♑	10 12 ♊
	8 3 ♉	*⊙ 20* ♒	9 14 ♉	3 13 ♒	11 5 ♋
25.	9 19 ♊	7 22 ♓	10 ⊙ 9 ♊	4 21 ♓	*⊙ 20* ♌
	11 1 ♋	8 8 ♈	11 3 ♋	6 2 ♈	12 8 ♍
	12 22 ♌	*⊙ 22* ♉	*⊙ 27·* ♌	7 13 ♉	year 12
	♀	9 10 ♊	12 21 ♍	8 29 ♊	♄
		10 ⊙ 8 ♋		10 13 ♋	*1 ⊙ 1* ♒
30.		11· 10 ♊		12 1 ♌	
		⊙ 12 ♋			

C₁ obverse.

I : Brugsch prints before most of the lines of this column three points the meaning of which is obscure to me (nothing is missing).

I,3 25 : sic! without ⊙

I,19 [1] 16 : sic! without ⊙; traces of 1 visible.

II,1 16 : sic! without ⊙

II,4 29 : sic! without ⊙

III,14/15 : traces of a sign are visible above the sign ≎ in line 15. These remains are just on the edge of the thickening in the middle of the tablet where pieces of plaster are missing.

IV,20 22 : sic! without ⊙. One would expect as preceding line 1 ⊙ 1 ♍, which perhaps is here omitted because of the same notice in the preceding year.

V,2 9 : sic! without ⊙.

V,13 5 : sic! without ⊙.

V,17 27 : sic! without ⊙.

V,26 8 : Brugsch reads 3 erroneously but gives 8 correctly in his drawing.

Stobart C_1

Reverse

Line	I	II	III	IV	V
1.	♃	6 13 ♒	8 9 ♓	1 ⊙ 1 ♍	6 17 ♓
	1 ⊙ 1 ♎	⊙ 30 ♓	9 4 ♈	2 13 ♎	8 17 ♈
	2 21 ♏	7 16 ♈	⊙ 28 ♉	4 ⊙ 1 ♏	9 6 ♉
	♂	8 10 ♉	10 23 ♊	5 13 [♐]	[⊙]21 ♊
5.	1 17 ♍	10 10 ♊	11 18 ♋	6 2[5 ♑]	[10] ⊙ 8 ♋
	3 5 ♎	⊙ 25 ♋	12 12 ♌	8 3 [♒]	[1]2 19 ♌
	4 23 ♏	11 10 ♌	☿	9 1[7 ♓]	[ye]ar 15
	6 7 ♐	12 9 ♍	*1* ⊙ 1 ♌	11 6 [♈]	♄
	7 29 ♑	⊙ 24 ♌	⊙ 16 ♍	♀	[1] ⊙ 1 ♓
10.	9 23 ♒	year 13	2 5 ♎	*1* ⊙ 1 [♍]	[9] 4 ♈
	♀	♄	⊙ 25 ♏	⊙ 23 [♎]	♃
	1 ⊙ 2 ♌	*1* ⊙ 1 ♒	3 14 ♐	2 19 [♏]	[1] ⊙ 1 ♑
	⊙ 29 ♍	7 24 ♓	5 19 ♑	3 13 [♐]	[4] 18 ♒
	2 23 ♎	♃	6 ⊙ 7 ♒	4 7 [♑]	[8] 5 ♓
15.	3 18 ♏	*1* ⊙ 1 ♏	⊙ 24 ♓	5 2 ♒	♂
	4 12 ♐	3 18 ♐	7 11 ♈	⊙ 25 ♓	*1* ⊙ 1 ♈
	5 7 ♑	♂	[9 14 ♉]	6 21 ♈	5 25 ♉
	6 1 ♒	*1* 5 ♒	[⊙ 29 ♊]	7 16 ♉	7 11 ♊
	⊙ 25 ♓	3 19 ♓	10 14 ♋	8 11 ♊	8 27 ♋
20.	7 19 ♈	5 2 ♈	11 4 ♌	9 9 ♋	10 21 ♌
	8 13 ♉	6 15 ♉	year 14	10 24 ♌	12 15 ♍
	9 7 ♊	8 3 ♊	♄	11 6 ♋	♀
	10 ⊙ 4 ♋	9 21 ♋	*1* ⊙ 1 ♒	☿	1 5 ♌
	⊙ 27 ♌	11 14 ♌	3 24 ♓	1 11 ♍	2 13 ♍
25.	11 21 ♍	epag. ⊙ 3 ♍	♃	⊙ 29 ♎	3 7 ♎
	12 16 ♎	♀	*1* ⊙ 1 ♐	2 19 ♏	4 3 ♏
	☿	1 6 ♏	4 5 ♑	3 9 ♐	⊙ 27 ♐
	1 ⊙ 1 ♍	2 9 ♐	8 15 ♒	⊙ 23 ♏	5 22 ♑
	2 12 ♎	6 9 ♑	10 26 ♑	4 2 ♐	6 16 ♒
30.	3 ⊙ 1 ♏	7 14 ♒	♂	5 13 ♑	7 11 ♓
	⊙ 20 ♐			⊙ 30 ♒	
	4 6 ♑				
	5 1 ♐				
	⊙ 19 ♑				

C_1 reverse.

I :   Brugsch prints before most of the lines in this column and in col. II,1 to 9 three points without visible reason.

I,5   17 :   sic! without ⊙.

I,10/11 :   one line missing, probably 12 20 ♑ (r.b.c.).

II,1 to 9 :   see note to col. I.

II,18   5 :   sic! without ⊙.

II,27   6 :   sic! without ⊙.

III,17 [9 14 ♉] :   traces of 9 and ♉ visible.

III,18 :   Brugsch did not restore this line which is however required first by the available space, secondly by the sequence of zodiacal signs, and thirdly by visible traces of ⊙.

IV,5   2[5 :   Brugsch 21 or 25, but 25 fits much better.

IV,7   1[7 :   traces only compatible with 1[2] or 1[7] and 17 fits better with the neighbouring lines, but T XVII, 10 has 13.

IV,10   [♍] :   traces visible.

IV,11   ⊙ 23 :   Brugsch "23 ou 25" because the text is destroyed just after the 3-sign. Four lines below we have however the sequence of dates 2→25→21 which speaks here more in favor of 1→23→19 than of 1→25→19.

IV,24   11 :   sic! without ⊙.

IV,29   2 :   sic! instead of 22.

V,4   [⊙] :   Brugsch incorrectly omits [ ].

V,10   [9] :   traces only compatible with 7 or 9; calculation shows that 9 is the only possible restoration.

V,13   [4] :   r.b.i. in agreement with calculation (♑ 28), and T III',18.

V,14   [8] 5 :   restoration according to T III',22.

V,23   5 :   sic! without ⊙.

Stobart C₂

Obverse

Line	I	II	III	IV	V
1.	8 5 ♈	♂	11 18 ♋	11 9 ♍	9 6 ♈
	☉ 29 ♉	[1] 23 ♎	12 7 ♌	12 18 ♎	10 17 ♉
	9 22 [♊]	[3] 7 ♏	☉ 22 ♍	☿	12 3 ♊
	10 16 [♋]	4 20 ♐	year 17	1 7 ♎	♀
5.	11 11 [♌]	6 2 ♑	♄	3 14 ♏	1 9 ♎
	12 5 [♍]	7 10 ♒	1 ☉ 1 ♈	4 1 ♐	☉ 29 ♍
	☉ 30 ♎	8 18 ♓	11 4 ♉	☉ 20 ♑	2 27 ♎
	☿	9 29 ♈	♃	5 9 ♒	4 12 ♏
	1 ☉ 2 ♍	11 8 ♉	1 ☉ 1 ♈	6 ☉ 4 ♑	5 11 ♐
10.	☉ 23 ♎	♀	9 1 ♉	☉ 20 ♒	6 6 ♑
	2 13 ♏	1 ☉ 19 ♏	♂	7 11 ♓	7 6 ♒
	3 ☉ 30 ♐	2 14 ♐	1 ☉ 9 ♊	☉ 27 ♈	☉ 26 ♓
	5 4 ♑	3 9 ♑	4 13 ♉	8 14 ♉	8 20 ♈
	☉ 21 ♒	4 6 ♒	6 12 ♊	9 7 ♊	9 15 ♉
15.	6 9 ♓	5 18 ♓	8 9 ♋	11 12 ♋	10 ☉ 9 ♊
	8 11 ♈	6 17 ♒	10 ☉ 1 ♌	☉ 28 ♌	11 3 ♋
	☉ 28 ♉	8 7 ♓	11 23 ♍	12 14 ♍	☉ 26 ♌
	9 14 ♊	9 16 ♈	♀	year 18	12 21 ♍
	10 ☉ 6 ♋	10 11 ♉	1 12 ♍	♄	☿
20.	12 13 ♌	11 6 ♊	2 10 ♎	1 ☉ 1 ♈	1 10 ♍
	epag. ☉ 4 ♍	12 1 ♋	3 4 ♏	7 ☉ 30 ♉	2 17 ♎
	year 16	☉ 25 ♌	☉ 28 ♐	♃	3 5 ♏
	♄	☿	4 22 ♑	1 ☉ 1 ♉	24 ♐
	1 21 ♈	1 16 ♎	5 16 ♒	9 ☉ 22 ♊	4 14 ♑
25.	2 1 ♓	2 13 ♏	6 10 ♓	♂	6 17 ♒
	6 7 ♈	☉ 15 ♎	7 5 ♈	1 4 ♎	7 5 ♓
	♃	3 19 ♏	☉ 29 ♉	2 19 ♏	☉ 21 ♈
	1 ☉ 1 ♒	4 9 ♐	8 24 ♊	3 28 ♐	8 19 ♉
	4 12 ♓	☉ 26 ♑	9 19 ♋	4 29 ♑	10 17 ♊
30.	8 15 ♈	5 14 ♒	10 14 ♌	6 17 ♒	11 1 ♋
		7 18 ♓		7 25 ♓	☉ 16 ♌
		8 4 ♈			12 6 ♍
		☉ 20 ♉			
		9 7 ♊			

C₂ obverse.

I,12 30 : reading certain (Brugsch "?").

I,24 21 : sic! instead of ☉ 1; obviously copyist error.

II,2 [1] 23 : sic! without ☉. Traces of 1 visible. No ☉ can be restored because 23 is written in ordinary number signs.

II,22 ☉ 25 : Brugsch erroneously translates "jour épagom. 5" but correctly draws ☉ 25.

II,24 16 : sic! without ☉.

III,19 12 : sic! without ☉. Instead of 12 one would expect 16 or 17; probably 2 is a misreading of a 7.

IV,4 7 : sic! without ☉.

IV,21 7 : Brugsch erroneously 10, because left part of the 7-sign effaced; reading 7 however certain and confirmed by calculation.

IV,24 22 : the sign for 2 almost closed circle (date-number sign).

IV,26 4 : sic! without ☉.

IV,29 4 29 : sic! instead of 5 7 (r.b.i. and controlled by calculation).

V,5 9 : sic! instead of ☉ 1.

V,11 6 : sic! instead of 1 or 2.

V,19/20 : one line omitted, probably 1 1 ♎; correspondingly line 20 should be ☉ 10 ♍.

V,20 10 : sic! without ☉.

V,23 24 : sic! without ☉.

Stobart C₂

Reverse

	I	II	III	IV	V
1.	year 19	6 11 ♒	7 19 ♈	12 5 ♎	♂
	♄	⊙ 27 ♓	8 12 ♉	♀	1 1 ♎
	1 ⊙ 1 ♉	7 14 ♈	9 7 ♊	1 ⊙ 2 ♏	⊙ 16 ♏
	♃	9 11 ♉	10 ⊙ 1 ♋	2 10 ♐	3 2 ♐
5.	1 ⊙ 1 ♊	10 ⊙ 6 ♊	⊙ 26 ♌	6 9 ♑	4 13 ♑
	10 15 ♋	⊙ 19 ♋	11 20 ♍	7 13 ♒	5 9 ♒
	♂	11 4 ♌	12 15 ♎	8 8 ♓	6 17 ♓
	1 ⊙ 1 ♊	year 1	♅	9 3 ♈	7 27 ♈
	⊙ 22 ♋	♄	1 ⊙ 1 ♌	⊙ 27 ♉	9 2 ♉
10.	9 12 ♌	1 ⊙ 1 ♉	⊙ 12 ♍	10 22 ♊	10 15 ♊
	11 3 ♍	9 13 ♊	2 3 ♎	11 16 ♋	12 ⊙ 30 ♋
	12 19 ♎	♃	22 ♏	12 10 ♌	♀
	♀	1 ⊙ 1 ♋	3 11 ♐	epag. ⊙ 1 ♍	1 1 ♍
	1 ⊙ 9 ♎	11 13 ♌	5 17 ♑	♅	⊙ 22 ♎
15.	2 3 ♏	♂	6 4 ♒	1 ⊙ 1 ♌	2 19 ♏
	⊙ 28 ♐	1 ⊙ 1 ♏	⊙ 20 ♓	⊙ 7 ♍	3 13 ♐
	3 22 ♑	3 11 ♐	7 8 [♈]	⊙ 26 ♎	4 7 ♑
	4 [1]7 ♒	4 20 ♑	9 [11 ♉]	2 16 ♏	5 2 ♒
	5 [12] ♓	5 26 ♒	⊙ 26 ♊	4 21 ♐	⊙ 26 ♓
20.	6 [7] ♈	7 3 ♓	10 11 ♋	5 10 ♑	6 21 ♈
	[7] 12 ♉	8 12 ♈	11 4 ♌	⊙ 28 ♒	7 16 ♉
	[9] 28 ♈	9 22 ♉	year 2	6 13 ♓	8 11 ♊
	[10 ⊙] 9 ♉	11 8 ♊	♄	8 16 ♈	9 12 ♋
	[11] 14 ♊	12 14 ♋	1 ⊙ 1 ♊	9 2 ♉	♅
25.	[1]2 13 ♋	♀	♃	⊙ 18 ♊	1 8 ♍
	♅	1 ⊙ 3 ♌	1 ⊙ 1 ♌	10 ⊙ 6 ♋	⊙ 28 ♎
	1 ⊙ 1 ♍	⊙ 28 ♍	11 16 ♍	12 23 ♌	2 11 ♏
	2 8 ♎	2 23 ♎	♂	year 3	3 9 ♎
	⊙ 27 ♏	3 17 ♏	1 ⊙ 1 ♋	♄	⊙ 13 ♏
30.	3 17 ♐	4 12 ♐	2 19 ♌	1 ⊙ 1 ♊	4 11 ♐
	5 ⊙ 20 ♑	5 6 ♑	10 14 ♍	11 1 ♋	5 1 ♑
		6 1 ♒		♃	⊙ 19 ♒
		⊙ 25 ♓		1 1 ♍	6 9 ♓

C₂ reverse.

I,1   year 19 :   ḥ³.t sp 10-t 9-t.

I,14   ♎ :   unusual form: a star is enclosed in the circle.

I,18 to 25 :   Brugsch indicates only slight damage at this place; obviously more has happened since then.

I,23   [10 ⊙] :   Brugsch gives 10 damaged and no ⊙, which according to the general rule (cf. p. 245) should be inserted.

II,7   11 :   Brugsch erroneously 12; copy correctly 11.

II,8   year 1 . . . :   see § 5, p. 244.

II,16   1 ⊙ 1 :   sic! instead of 1 ⊙ 30 or 2 ⊙ 1. The first correction is epigraphically better because of the similarity of ⊙ 30 and ⊙ 1; cf. the analogous correction in V,11.

II,26   ⊙ 3 :   Brugsch "jour 3 (?)." I do not see the reason for the "?".

III,12   22 :   sic! without ⊙.

IV,33   1 1 :   sic! without ⊙.

V,2   1 1 :   sic! without ⊙.

V,11   ⊙ 30 :   sic! instead of ⊙ 1; cf. II,16.

V,25   8 :   sic! without ⊙.

Stobart E

Obverse

	I			II			III			IV			V		
1.	☉ 28	♉			♂		☉ 29	♈		4	[1	♏]		♄		
	10	22	♊	1 ☉ 1	[♓]		8	15	♉	☉ [26	♐]		1 ☉ 1	♎		
	11	17	⊛	4	[6	♈]	9	6	♊	5	20	[♑]	[5]	11	♏	
	12	12	♌	6	5	♉	11	14	⊛	6	1[5	≈]	[6]	26	♎	
5.	☿			7	21	♊	☉ 29	♌		7	[9	♓]	♃			
	1 ☉ 1	♍		9	7	⊛	12	16	♍	8	[2	♈]	[1] ☉ 1	⊛		
	[☉] 18	≈		10 ☉ 30	♌		year	12		☉ [27	♉]		[11]	1	♌	
	[2] 8	(♍)		12	[1]8	♍	♄			9	[21	♊]	♂			
	[☉ 2]5	≈		♀			1 ☉ 1	♎		10	1[5	⊛]	[1 ☉] 1	♉		
10.	[3]	19	♏	1 ☉ 1	♍		♃			11	10	[♌]	[5]	11	♊	
	4	10	♐	☉ 25	≈		1 ☉ [1	♊]		[12]	4	≈	8	18	⊛	
	☉ 28	♑		2	19	♏	10 ☉ 3	⊛		[☉ 28]	≈		10	10	♌	
	5	16	≈	3	13	♐	♂			☿			12	2	♍	
	7	18	♓	4	7	♑	////////////////////////			1 ☉ 1	♍		♀			
15.	8	8	♈	5 ☉ [1]	≈		////////////////////////			☉ 1[2]	≈		1 ☉ 1	≈		
	☉ 2[8	♉]		☉ 26	♓		1 ☉ 1	♍		☉[2]3	♍		☉ 18	♏		
	9	9	♊	6	21	♈	[2]	1	≈	2	1[7]	≈	2	13	♐	
	11	17	⊛	7	[16]	♉	3	17	♏	3	5	♏	3	8	♑	
	12	7	♌	8	11	♊	4	29	♐	[☉ 23	♐]		4	8	≈	
20.	☉ 25	♍		9	12	⊛	6	11	♑	[4	10	♑]	[8]	10	♓	
	year	11		☿			7	19	≈	6	1[3	≈]	9	17	♈	
	♄			1 ☉ 1	♍		8	27	♓	7	5	[♓]	10	12	♉	
	1 ☉ 1	♍		☉ 8	≈		10 ☉ 8	[♈]		☉ 21	[♈]		11	7	♊	
	2	1	≈	3	17	♏	1[1 20]	♉		[8 9]	♉		12	2	[⊛]	
25.	♃			4	4	♐	[♀]			[10 1]6	♊		☉ 27	♌		
	1 ☉ 1	♉		☉ 18	♑		1 ☉ [15	♌]		11 ☉ 1	⊛		☿			
	9	13	♊	5 ☉ 10	≈		2	11	♍	☉ 18	♌					
				6	2	♑	3	6	≈	12	24	♍				
				☉ [18]	≈					year	13					
30.				7	[13]	♓										

E obverse.

I,1 ♉ : traces of a sign are visible (overlooked by Brugsch) after the usual sign of ♉ which may be restored to the sign "star" which occurs in rev. V,1 after ♈.

I,7 [☉] : still visible on Brugsch's copy.

I,8 (♍) : Brugsch incorrectly "detruit," but 8 clearly visible and the zodiacal sign perhaps omitted.

I,9 [☉ 2]5 : r.b.i.; only second half of 5 visible.

I,10 [3] : traces visible.

I,16 2[8] : Brugsch 24, but there is space enough for a second 4-sign which is required by interpolation.

II,3 [6] : Brugsch gives 6 without ( ) in his translation, but no sign in his copy.

II,12 19 : Brugsch erroneously 9; copy correctly 19.

II,15 [1] : Brugsch in his copy and translation 4, but only weak traces are visible, which also fit with 1.

II,18 [16] : sic according to Brugsch; traces also compatible with 15.

II,19 11 : certain.

II,29 [18] : Brugsch still has the 10-sign.

II,30 : this line erroneously omitted by Brugsch in copy and translation.

III,4 14 : certain.

III,14/15 : thickening in the middle of the tablet; no text missing.

III,16 to 28 : disregarded by Brugsch.

IV,1 [1 : traces of a stroke visible, which could only belong to a 9, which is excluded by calculation.

IV,2 ☉ [26 : traces of a sign visible which would be best restored to 10 but 20 necessary by interpolation.

IV,3 20 : additional units possible but not required by interpolation.

IV,4 to 10 : these lines only recorded by Brugsch.

IV,4 1[5 : traces of 5 visible.

IV,5 [9 : certain because of traces.

IV,6 [2 : traces visible.

IV,7 ☉ [27 : Brugsch "jour 19," but his copy shows no 9 and the other traces fit as well to 10 as to 20.

IV,8 9 : omitted by Brugsch.

IV,12 28] : traces of 8 visible.

IV,15 1[2] : weak traces of 2 visible.

IV,16 [2]3 : traces of 20 visible.

IV,17 1[7] : traces of 7 visible.

IV,26 ☉ 1 : reading very uncertain.

IV,28 24 : reading of 4 very uncertain.

V,2 1 ☉ 1 : Brugsch: "(1) jour 1" without reason for ( ).

V,3 [5] : r.b.c.; the planet is so near to the stationary point that [4] would also be possible.

V,4 [6] : r.b.c.; the planet is so near to the stationary point that 7 would also be possible.

V,7 to 26 : disregarded by Brugsch.

V,26 ☿ : repeated in rev. I,1.

V,27 : empty.

Stobart E

Reverse	I	II	III	IV	V
1.	☿	♀	♃	☉ 25 ♑	6 6 ♈
	1 ☉ 1 ♍	1 ☉ 1 ♌	1 ☉ 1 ♍	5 28 ♒	7 14 ♉
	2 9 ♎	☉ 16 ♍	♂	6 14 ♓	9 10 ♈
	☉ 29 ♏	2 10 ♎	1 ☉ 1 ♊	8 17 ♈	10 ☉ 2 ♉
5.	3 19 ♐	3 5 ♏	2 23 ♋	9 4 ♉	11 16 ♊
	4 9 ♑	☉ 29 ♐	4 24 ♊	☉ 20 ♊	12 12 ♋
	☉ 20 ♐	4 21 ♑	7 27 ♋	10 ☉ 7 ♋	☿
	5 14 ♑	5 15 ♒	9 21 ♌	12 19 ♌	1 ☉ 3 ♍
	6 9 ♒	6 9 ♓	11 12 ♍	year 16	☉ 24 ♎
10.	☉ 28 ♓	7 4 ♈	12 28 ♎	♄	2 11 ♏
	7 15 ♈	☉ 28 ♉	♀	1 ☉ 1 ♏	4 16 ♐
	8 27 ♉	8 23 ♊	1 ☉ 1 ♎	4 22 ♐	5 3 ♑
	10 ☉ 7 ♊	9 18 ♋	☉ 23 ♍	9 4 ♏	☉ 20 ♒
	☉ 21 ♋	10 13 ♌	3 ☉ 1 ♎	♃	6 8 ♓
15.	11 6 ♌	11 9 ♍	4 14 ♏	1 ☉ 1 ♎	7 8 ♒
	year 14	12 21 ♎	5 11 ♐	♂	15 ♓
	♄	☿	[6] 5 [♑]	1 ☉ 1 ♎	8 9 ♈
	1 ☉ 1 ♎	1 ☉ 1 ♌	☉ 30 ♒	2 9 ♏	☉ 26 ♉
	2 1 ♏	☉ 12 ♍	7 24 ♓	3 18 ♐	9 12 ♊
20.	♃	2 2 ♎	8 19 ♈	4 27 ♑	10 ☉ 5 ♋
	1 ☉ 1 ♌	☉ 24 ♏	9 14 ♉	6 4 ♒	12 12 ♌
	12 4 ♍	3 13 ♐	10 ☉ 8 ♊	7 11 ♓	☉ 30 ♍
	♂	5 18 ♑	11 ☉ 4 ♋	8 22 ♈	year 17
	1 ☉ 1 ♍	6 5 ♒	☉ 26 ♌	10 ☉ 3 ♉	♄
25.	☉ 12 ♎	☉ 24 ♓	12 20 ♍	11 19 ♊	1 ☉ 24 ♐
	2 27 ♏	7 10 ♈	☿	♀	♃
	4 7 ♐	9 12 ♉	1 ☉ 1 ♌	1 8 ♎	1 ☉ 1 ♎
	5 17 ♑	☉ 27 ♊	☉ 8 ♍	2 8 ♏	2 2 ♏
	6 25 ♒	10 12 ♋	☉ 28 ♎	☉ 27 ♐	6 29 ♐
30.	8 3 ♓	11 4 ♌	2 17 ♏	3 22 ♑	8 10 ♏
	9 14 ♈	year 15	4 6 ♐	4 16 ♒	
	10 25 ♉	♄		5 11 ♓	
	12 11 ♊	1 ☉ 1 ♏			

E reverse.

I,12 8 27 ♉: sic! probably by contraction of the following three lines

8	3	♉
☉	27	♈
9	19	♉

I,13 ☉: omitted in Brugsch's translation.

II,33: Brugsch ". . ." without reason.

IV,2 28: sic! instead of 18.

IV,6 20: Brugsch's translation erroneously 10, copy correctly 20.

IV,8 19: Brugsch's translation erroneously 9, copy correctly 19.

IV,27 8: sic! without ☉.

IV,28 8: sic! instead of 3; obviously caused by the preceding 8.

V,1 ♈: the sign ♈ followed by the sign "star" (cf. obv. I,1).

V,16 15: sic! without ☉.

V,30 10: Brugsch reads 14 but the sign which he took as 4 probably belongs to the sign of ♏.

§ 4. Commentary

In the following we shall discuss two essentially different problems: first of all we shall give a detailed explanation of the texts and show how the recorded numbers describe the planetary positions; these questions will be treated in the "astronomical" part of this commentary. The second problem is of strictly historical character: are those texts of purely Egyptian origin or written under Greek influence? This leads to an investigation of the ancient reports on Egyptian astronomy, given in the "historical" section.

The combination of the results of these two groups of investigations should finally give us a picture of the character of our texts. Unfortunately however the material available is not sufficient to enable us to arrive at comparative results. It will be seen that we are able to obtain an almost complete understanding of the astronomical structure of the texts (except for a small number of dates in P concerning Saturn); on the contrary the result of the historical discussion turns out to be very negative. In spite of a large number of remarks about Egyptian astronomy in Greek and Roman literature, the existing knowledge about an Egyptian theory of the planets seems to have been practically negligible. There originated however an arbitrary historic construction based on a fictitious conception of the beginnings of mathematics which today can easily be proved to be wrong. We will therefore at the end be compelled to give only some more or less plausible arguments about the general history of our texts.

A. ASTRONOMICAL COMMENTARY.

1. General Remarks.

Both P and S give positions from year to year of the five planets in the ecliptic according to the following order:

Saturn ($\hbar$)   Jupiter ($\mathcal{2}\mathcal{L}$)   Mars ($\vec{\sigma}$)
Venus (φ)   Mercury ($\mathcal{Y}$).

The positions are indicated in the order: month (from 1 to 12 and epagomenae), day (from 1 to 30) and zodiacal sign. It is at first glance very plausible that these combinations of dates and signs mean precisely the moment of the entrance of the planet into the sign, of course sometimes retrograde, as is indicated by the reversed order of the signs.[20] The correctness of this assumption is proved by astronomical calculation in the overwhelming majority of all cases. Some exceptions will be treated in their proper places.

The outlines of the structure of our texts are thus established. Before we go into detail we must discuss two important preliminary questions: what is the calendar-system used by the texts and how are the

zodiacal signs situated with respect to the exact vernal point?

2. The Calendar.

As is well known, two different calendars existed simultaneously in Egypt during Roman times: the old "Egyptian calendar," which consistently uses "years" of 365 days and the "Alexandrian calendar" which inserts at the end of every fourth year one more day, but otherwise uses the same names and lengths for the months as the Egyptian calendar. The more remote from the date of introduction of the "Alexandrian" practice of intercalation the more the difference between the two calendars increased. In neither P nor S does a sixth intercalary day occur, which would be a definite sign of the use of the Alexandrian calendar; we must therefore recur to astronomical methods in order to decide whether the recorded dates are to be understood in the "Egyptian" or in the "Alexandrian" manner.

Our procedure is very simple. The number of days elapsed from the date of the introduction of the Alexandrian calendar to some date called "month a day b" is smaller in the Egyptian than in the Alexandrian calendar. Hence, if we calculate the position of a planet at the date "month a day b," then calculation with Alexandrian dates will result in a greater value of longitude [21] than with Egyptian dates, increasing in proportion to the remoteness from the date of coincidence.[22] In other words: the *same* value of longitude (say entrance into a certain sign) appears on an *earlier* date, if these dates belong to the Alexandrian calendar. Therefore, if we represent both sets of calculated positions in a time-longitude-diagram, the curve connecting the points calculated by using the Egyptian calendar will deviate more and more to the right of the corresponding Alexandrian curve. If we finally introduce into this diagram the points corresponding to dates and longitudes recorded in the text we can immediately decide which calendar has been used: only one of the calculated curves can coincide with or at least run parallel to the curve determined by the text-positions, while the other deviates uniformly from this pair.

In order to apply this method we consider the movement of Venus. This planet moves fast enough to make even small differences in time influential on its position and, on the other hand, so regularly that a good agreement between text and calculation can

---

[20] This assumption was already made by Brugsch [1] p. 22.

[21] Strictly speaking heliocentric longitudes only, but if we restrict ourselves to the direct part of the orbit the same is true for geocentric longitudes.

[22] We are using as date of coincidence the epoch on which Schram's tables are based: Augustus 1 Thoth 1 = − 29 August 29. Concerning the question of the actual first intercalation, see Ginzel, Chron I p. 226 ff., which, however, does not affect our calculation.

be expected.[23] We start therefore with the position of Venus at the most remote time available which is the year Hadrian 16 ($\approx$ 132 A.D.). The two calculated curves differ by 40 days corresponding to the 160 years elapsed since the introduction of the Alexandrian calendar (see Pl. 9 at the end); if we now compare the positions given in the text with these two curves the agreement between text and Alexandrian curve is so obvious that this alone would speak strongly in favor for this calendar. The definite proof of its use lies in the fact that the agreement between text and Alexandrian curve is still the same

3. The Origin of the Zodiac.

After having determined the meaning of the dates recorded in P and S we must now discuss the references to the zodiac. Experience from all the calculations carried out in order to compare the text positions with the facts shows that the text positions are on the average a few degrees in advance of the calculated positions, with the single exception of Mercury. The average numbers [26] (in degrees) of

$$\Delta = \text{calculation} - \text{text}$$

are as follows:

text		year	☿	♀	♂	♃	♄	year
P	−16	Augustus 14	+1	−4	−6	−3	−4	Augustus 14
	+10	Augustus 40	+1½	−1	−5			to 40
S	+70	Vespasian 3	+3½	−3	−1½	−1	−4	Vespasian 3 to
	+131	Hadrian 16	+4	−2	−2½			Hadrian 16

at the beginning of the available interval (Vespasian 4 $\approx$ 71 A.D.) as Pl. 8 shows.[24]

The situation is much less favorable in the case of P because the youngest part belongs to Augustus 40 ($\approx$ 10 A.D.) which corresponds to only about 10 days difference between the two calendars. The calculation of the positions of Venus during the years Augustus 14 to 19 under the assumption of the Egyptian calendar gives a curve which lies on the average about 4 degrees below the text-positions (Pl. 6 shows this for the year Augustus 14). The corresponding Alexandrian curve lies of course a little closer to the text. However, at the end of the interval (Augustus 40; cf. Pl. 7), the Egyptian curve coincides very closely with the text, although still lying about one degree below it, while the Alexandrian curve lies about 8 degrees above the text (see Pl. 7). This shows that the curve of the Egyptian calendar seems to approach the text somewhat (from about −4 to −1), but the Alexandrian curve agrees with the text at the beginning (with a little negative deviation) and deviates considerably (positively) at the end, which indicates clearly that the Egyptian and not the Alexandrian calendar was used in P.[25]

I am not able to explain the positive values of Δ in the case of Mercury. The consistent negative Δ in all other cases can easily be understood as the result of a little difference in the definition of the longitudes. The modern calculations are of course based on the true vernal-point as origin. If we assume that longitudes were counted from some fixed star a difference of a few degrees would be not at all surprising.

However, the assumption of some fixed star as origin of the zodiac introduces further consequences. The calculations by which we check the records of the texts are made by using "longitudes" counted from the actual vernalpoint at this time. Therefore the "longitude" λ of a fixed star, supposed to be the boundary of a zodiacal sign, will increase as a function of time by 1° in about 72 years, because of the precession of the equinoxes. If the texts give the entrance of a planet into a zodiacal sign twice, first in the year Augustus 14 (= 17 B.C.) and secondly in the year Hadrian 17 (= 132 A.D.), then the calculated longitudes must be about 2° greater in the second instance than in the first. But the differences

---

[23] Mercury is less adequate for our purpose because of the larger irregularity of its movement and the relative higher frequency of retrograde parts in its orbit.

[24] This agreement with the Alexandrian calendar has already been found by Ellis and Airy (cf. Brugsch [1] p. 63).

[25] Due to the fact that P is based on the Egyptian calendar, S to the contrary on the Alexandrian calendar, no conclusions can be derived as to the use of those calendars in general. The Demotic Pap. Carlsberg 9, written after 144 A.D. (Neugebauer—Volten [1] p. 383), or Ptolemy, in the Almagest, uses the Egyptian calendar consistently. The same is still true for the commentaries on the Almagest; Pappus gives e.g. the date of the sun-

eclipse of Oct. 18, 320 as "Era Nabonassar 1068 Tybi 17," which assumes the Egyptian calendar (Pappus, In Almag. VI, ed. Rome p. 180, 9 f.). During these centuries both calendars were obviously used simultaneously in Egypt.

[26] There is not much sense in trying to give an exact definition of what an "average" means here. I have, of course, disregarded obvious errors in the texts, and also the retrograde parts because of the special kind of deviations here. What I attempted to do is only to represent in the diagrams shown on Pls. 1 to 15 the mean distance between the "linear" parts of calculated curve and text. The calculation itself admits an error of about 1 degree but it is senseless to use more accurate methods because the dates are given only in days, which involves an error margin of the same order of magnitude.

between calculation and text, given above (p. 230), are calculated disregarding precession by simply calling $\lambda = 0$ the beginning of ♈, $\lambda = 30$ the beginning of ♉, etc. We must therefore expect that the values of Δ increase (algebraically) by about 2° from the beginning of P to the end of S, or about 1° during S alone, if we wish to avoid comparing different texts. It seems to me, that the numbers derived from p. 230 agree sufficiently well with this consequence:

deviations in the interior of our interval must therefore be of periodic character. This assumption is in itself extremely unlikely because these periods must fit not only into the purely accidental intervals of 25 years in P and 60 years in S but be common for all three planets Mercury, Venus and Mars, which is astronomically senseless. It seemed to me therefore fully sufficient to confine the investigation of the recorded positions of those three planets to the following:

increase of Δ	☿	♀	♂	♃	♄	expected
from −16 to +131	+3	+2	+3½	+2	0	+2
from +70 to +131	+½	+1	−1½	— [27]	— [27]	+1

This makes it very probable that both texts are using a fixed origin for the division of the zodiac into twelve signs, disregarding precession. If this be true, then the list p. 230 shows the origin of this fixed zodiac at the beginning of the Augustian time to be about four degrees in advance of the vernal-point ($\lambda = 356°$).[28]

4. Mercury, Venus, Mars.

The leading principle in our text of indicating the dates of entrance of a planet into a zodiacal sign causes considerable difference in the treatment of Jupiter and Saturn on one side and Mars, Venus and Mercury on the other. The movement of the three planets nearest to the sun is fast enough to give sufficiently many points in our diagram-representation (Pls. 1 to 11) to interpolate a continuous curve, comparable to the curve representing the actual movement found by calculation. Jupiter and Saturn, however, move so slowly that the time difference between successive entrances into a sign becomes much too large to produce a sufficient number of points for graphic interpolation. Here, therefore, it was necessary to compute the actual position for every date given in the texts. In the case of the three interior planets, however, a simpler method of checking the text could be applied. The comparison between text and calculation at both ends of P and of S showed no kind of systematic deviation.[29] The only possible systematic

all recorded positions were plotted in a time-longitude diagram and considered to be correct if the curves obtained showed the same regular type as the parts at the beginning and at the end, actually compared with calculation. And indeed, where some doubtful cases were controlled by calculation, the same agreement between the regular parts of the diagram and the actual positions, established in the beginning and ending of the two texts, always appeared. There can therefore be no serious doubt that, in general, the texts agree very closely with the actual movement.

The texts disregard, however, all latitudes and the same was correspondingly done in our calculations. In order to show what this means for the actually observable positions of a planet, the orbit of Venus was calculated exactly both in longitude and latitude for the period Trajan 17, X 14 (= 114 A.D. June 8) to Trajan 18, IV 9 (= 114 A.D. December 8) for ten-day intervals. The orbit thus obtained is shown on plate 16; the small circles indicate the positions of Venus at ten-day intervals, the crosses the positions at the days mentioned in the text.[30] In the retrograde part Venus is invisible from A to B, but, in spite of this fact, the text gives a date which practically coincides with the inferior conjunction. This corresponds to the rule, obvious in our texts, that the entrance into a sign has to be indicated whether it is visible or not.

A second general fact can be observed for the retrograde parts of the orbits, namely that the corresponding waves in the diagrams generally have too small amplitudes. Mercury in Vespasian 3, VIII (Pl. 4) and Hadrian 16, VII (Pl. 5), Venus in Hadrian 16, IX/X (Pl. 9), Mars in Hadrian 14, V (Pl. 11) are

---

[27] In the case of Jupiter and Saturn there are not sufficiently many dates recorded to give an average deviation for shorter intervals.

[28] It may be remarked that Kugler discovered that the Babylonian planetary texts, which belong to the two last centuries B.C., use a vernalpoint about five degrees in advance of the true vernalpoint (cf. e.g. Kugler SSB I p. 121, p. 173 and SSB II p. 513 ff.) but this correspondence might be purely accidental. There is no reason whatsoever to assume that Babylonian astronomy took into account the precession of the equinoxes. Schnabel's paper [1] can be disproved in every detail.

[29] When speaking about "systematic deviations" I no longer consider the displacement of the calculated curve with respect to

the text-curve caused by the different origins and by precession, discussed in the preceding section.

[30] The boundary lines of the zodiacal signs are placed 2 degrees before the corresponding longitudes, which refer to the true vernalpoint. According to the discussion above this may be a fairly correct determination of the boundary lines intended in the text.

typical examples. This is less surprising in the case of Venus and Mercury, where a large portion of the retrograde movement is invisible, but should not be expected in the case of an exterior planet for which the retrograde movement is fully visible.[31]

The largest irregularities in the description of the path of the first three planets appear in the case of Mercury, because of the frequency of retrograde movements and the correspondingly large irregular variation in the apparent velocity. But even so the agreement is in general better than one would expect offhand (cf. Pls. 1 to 5). How excellent the results can be in the case of Venus and Mars is shown on plates 8 and 9.

There remains the discussion of a little peculiarity of S. In the Stobart tablets the majority of the groups referring to the same planet begins with the date "first month, day 1" (i.e. "New Year's Day") and it is obvious that in many of those instances no definite position can be meant. Examples are:

☿			
Trajan	9	XII	10	♌
		XII	28	♍
Trajan	10	I	1	♍
		I	11	♎

♀			
Hadrian	13	XII	2	♋
		XII	27	♌
Hardian	14	I	1	♌
		I	16	♍

♂			
Trajan	18	X	17	♉
		XII	3	♊
Trajan	19	I	1	♊
		I	22	♋

In all these cases the same zodiacal sign corresponds both to New Year's Day and to the preceding date, which indicates clearly enough that at the New Year's Day only a position somewhere inside the sign is meant. There exist, on the other hand, cases where a New Year's Day is "significant," i.e. where the planet crosses the frontier of a sign on this day:

☿			
Vespasian	4	XII	7	♌
Vespasian	5	I	1	♍
		I	20	♎

---

[31] Mars is therefore called "the retrograde moving star" (cf. Brugsch, Thes. I p. 65 and 69).

♀			
Trajan	13	XII	12	♌
Trajan	14	I	1	♍
		I	23	♎

No repetition of signs occurs here.

The simple explanation of the occurrence of the non-significant New Year's Days lies in the fact that the scribe of S intended to mention for every year and each planet the sign inside which the planet stands in the first month of the year. When, as in many cases, the planet did not cross the boundary line of a sign during the first month, he then simply took the New Year's Day. According to this custom he sometimes introduced the New Year's Day even where some other significant date fell in the first month; in the majority of cases, however, he was satisfied if at least one date appeared in the first month.

5. Jupiter and Saturn.

At first glance the meaning of the dates given for ♃ and ♄ seems to be quite different from the simple scheme of the three other planets. We take, for example, the first ten dates for ♄ in P:

Augustus	15	3	21	♐
	16	7	1	[♑]
		10	5	[♐]
	17	3	30	♑
	18	4	28	♑
	19	6	6	♒
	20	1	25	♒
	21	8	14	♓
	22	5	2[7]	♓
	23	[12]	5	♓

If only the moments of the entrance of a planet into a sign were recorded, then pairs or even triplets of the same sign can never occur. We must, on the other hand, realize that we should at any rate expect some kind of new rules in the case of such slowly moving planets as Jupiter and Saturn, because there must be years in which a planet does not leave a zodiacal sign. It follows therefore from the mere fact that an entry for ♃ and ♄ is given in *every* year that a new notation must have been adopted.

These remarks determine our following procedure. We must separate the dates given in the texts into two classes: one which refers to the entrance (direct or retrograde) into a sign, the other including the remaining dates. We shall first attempt it in the simplest case, that is, of Jupiter, and in the better text, that is, S.

The dates of the first class must fulfill the requirement that two equal signs never follow each other, a condition which in practice is almost sufficient to eliminate the rest; control by modern calculation finally excludes every possible doubt. This gives the following list of "significant" dates (retrograde entrance marked by ↓):

♃

	year	month	day	sign	position	calculation −text
A	Vespasian 4	[1]1	24	♎ ♎	3	+ 3
	5	4	6	♏ ♎	27	− 3
		9	16	♎ ♏	1 ↓	+ 1
		12	16	♏ ♏	0	0
	6	4	21	♐ ♏	26	− 4
	7	5	12	♑ ♐	27	− 3
	8	5	[30]	♒ ♑	28	− 2
	9	6	14	♓ ♓	1	+ 1
	10	6	23	♈ ♓	29	− 1
		11	1[1]	♉ ♈	28	− 2
C₁	Trajan 9	12	28	♍ ♌	29	− 1
	11	1	22	♎ ♎	0	0
	12	2	21	♏ ♎	29	− 1
	13	3	18	♐ ♐	1	+ 1
	14	4	5	♑ ♑	0	0
		8	15	♒ ♑	25	− 5
		10	26	♑ ♑	26 ↓	− 4
	15	[4]	18	♒ ♑	28	[− 2]
		[8]	5	♓ ♒	24	[− 6]
C₂	16	1	1	♒ ♒	29 ↓	− 1
		4	12	♓ ♒	28	− 2
		8	15	♈ ♓	24	− 6
	17	9	1	♉ ♈	25	− 5
	18	9	22	♊ ♉	28	− 2
	19	10	15	♋ ♊	29	− 1
	Hadrian 1	11	13	♌ ♋	29	− 1
	2	11	16	♍ ♌	24	− 6
E	11	9	13	♊ ♊	0	0
	12	10	3	♋ ♊	29	− 1
	13	[11]	1	♌ ♌	0	[0]
	14	12	4	♍ ♍	1	+ 1
	16	1	1	♎ ♎	2	+ 2
	17	2	2	♏ ♏	3	+ 3
		6	29	♐ ♏	29	− 1
		8	10	♏ ♏	29 ↓	− 1

This list shows clearly the correctness of our assumption that it is possible to find a sequence of "significant" dates which indicate as in the case of the three first planets the complete list of crossings of the boundary lines. Furthermore, there exist years without such crossings, namely Trajan 10 and Hadrian 15, Jupiter being in both cases inside ♍.

If we now consider the dates not yet contained in the preceding list, we find the very simple result that those dates are without exception New Year's Days in consecutive years, including the years Trajan 10 and Hadrian 15 just mentioned. This shows that the dates of the text are simply the combination of all "significant" dates and all New Year's Days.[32] Plate 13 gives the graphic representation of this situation; the points on the vertical lines are the New Year's Days and the points near to the horizontal boundary lines correspond to the "significant" dates.[33]

Before we extend our discussion to P, we should remark that the list given above shows that the agreement between significant dates and calculation is a very good one. The positions in A deviate from calculation by − 1°, in C₁ + C₂ by − 2° and perhaps slightly positively in E (common average about − 1°). The only disagreement seems to appear in the first listed year (Vespasian 4) where we should expect two more dates, because, according to calculation, ♃ crosses the line ♎ 0° in this year three times.

We now turn to P (cf. Pl. 12). The list of "significant" dates is in the first column of page 234 giving exactly as in S the complete movement.[34] The agreement with the calculation is not quite as good as in S, especially if we consider single values which go down to − 7°, the average, however, being only − 3.

Very different however is the situation in P as far as the remaining non-significant dates are concerned. P never uses "New Year Days" (except when they are significant[35]). There are, however, two years where ♃ does not leave the sign ♌, namely year 28 and 40, and the only two dates from the text not yet listed above refer to those two years:

Augustus 28 | IV [3] ♌ ‖ calculation: ♌ 11
Augustus 40 | IV 16 ♌ ‖ calculation: ♌ 15 ↓.

The purpose in giving these dates was obviously to avoid a gap in the sequence of years. The calculated positions ♌ 11[36] and ♌ 15 indicate that the dates

[32] There can be, of course, New-Year-Days which are by chance "significant." This is the case in the above given list in the years Trajan 16 and Hadrian 16.

[33] The plates 12 to 15 contain only *calculated* positions. The text-positions lie in all cases on the same abscissa and in the case of the "significant" dates on the nearest boundary line of the zodiacal signs. There is obviously no definite position attached to the non-significant dates. The continuous curves do not pretend to give the movement exactly but show merely the distribution of direct and retrograde parts in the apparent orbit.

[34] Except one omission of a retrograde entrance in the year 18.

[35] This occurs only once in the whole text: ♀ year 14. For the case ♄ year 37 see p. 235 below.

[36] This is almost exactly a stationary point; therefore the uncertainty in the reading of the day-number does not affect the position.

♃

	text			calculation	
Augustus year	month	day	sign	position	calculation − text
15	6	8	♊	♊ 25 ↓	− 5
	7	27	⊗	♊ 26	− 4
16	1	[15]	♌	⊗ 29	[− 1]
	7	[5]	⊗	⊗ 29 ↓	[− 1]
	8	[30]	♌	⊗ 28	[− 2]
17	2	20	♍	♌ 29	− 1
	8	5	♌	♌ 29 ↓	− 1
	9	29	♍	♌ 28	− 2
18	3	20	<≈>	♍ 29	− 1
	<9	9>	♍	♍ 28 ↓	[− 2]
	1[1]	9	≈	♍ 29	[− 1]
19	4	22	♏	♏ 0	0
	10	12	≈	≈ 29 ↓	− 1
	11	27	♏	≈ 28	− 2
20	[4]	23	♐	♏ 27	[− 3]
21	5	18	♑	♐ 27	− 3
22	6	3	. ≈	♑ 27	− 3
23	[6]	8	♓	≈ 27	[− 3]
	11	12	♈	♓ 25	− 5
24	1	8	♓	♓ 24 ↓	− 6
	6	5	♈	♓ 24	− 6
	11	5	♉	♈ 27	− 3
25	4	1	♈	♈ 23 ↓	− 7
	6	3	♉	♈ 24	− 6
	[11	1]8	♊	♉ 27	[− 3]
26	2	29	♉	♊ 5 ↓	+ 5
	6	7	♊	♉ 27	− 3
	12	1	⊗	♊ 27	− 3
27	12	29	♌	⊗ 28	− 2
29	1	29	♍	♌ 27	− 3
30	[3]	4	≈	♍ 29	[− 1]
31	3	9	♏	≈ 24	− 6
32	4	18	♐	♏ 27	− 3
- - - -	- - - -	- - - -	- - - -	- - - - -	- - - -
35	6	4	♓	♓ 0	0
	10	12	♈	♓ 26	− 4
36	[1	2]1	♓	♓ 26 ↓	[− 4]
	[6	5]	♈	♓ 28	[− 2]
	[10	25]	♉	♈ 27	[− 3]
37	[10	2]4	♊	♉ 25	[− 5]
38	1[1]	14	⊗	♊ 25	[− 5]
39	12	12	♌	⊗ 25	− 5
41	1	12	♍	♌ 27	− 3

were chosen in such a way that ♃ then stands exactly in the center of the sign which it does not leave during this year. However, two instances are not sufficient proof for this hypothesis and the corresponding problem in the case of ♄ does not support this assumption as will be shown (p. 235).

Applying the same methods to Saturn we find essentially the same results in S (cf. Pl. 15) with the only difference that more years are missing in the list of the significant dates [37] because of the very slow movement of Saturn.[38]

♄

		text				calculation	
	year	month	day	sign	position		calculation − text
A	Vespasian 5	5	5	♐	♏	28	− 2
		9	25	♏	♏	29 ↓	− 1
	6	2	7	♐	♏	28	− 2
	8	4	[18]	♑	♐	25	[− 5]
	10	7	1	≈	♑	25	− 5
		12	[3]	♑	♑	25 ↓	[− 5]
C₁	Trajan 9	3	5	♑	♐	27	− 3
		11	5	14	≈	♑	25	− 5
	13	7	24	♓	≈	24	− 6
	14	3	24	♓	≈	24 ↕	− 6
	15	9	4	♈	♓	22	− 8
C₂	16	2	1	♓	♓	23 ↓	− 7
		6	7	♈	♓	23	− 7
	17	11	4	♉	♈	23	− 7
	18	1	1	♈	♈	24 ↓	− 6
		7	30	♉	♈	25	− 5
E	Hadrian 1	9	13	♊	♉	25	− 5
	3	11	1	⊗	♊	25	− 5.
	11	2	1	≈	♍	29	− 1
	13	[5]	11	♏	≈	29	[− 1]
		[6]	26	≈	≈	29 ↓	[− 1]
	14	2	1	♏	≈	28	− 2
	16	4	22	♐	♏	28	− 2
		9	4	♏	♏	2	+ 2
	17	1	24	♐	♏	29	− 1

[37] There are a few cases where ♄ crosses the boundary line of a sign according to calculation but which are omitted in the text. Calculation gives e.g. for Vespasian 9, I 1 the place ♐ 29 and similar cases are Trajan 12 and Hadrian 11 (cf. Pl. 15). The reason for omitting these cases is the above mentioned fact (p. 231) that the longitudes given in the text are advanced about four degrees compared with calculation.

[38] The sign ↕ means: stationary point between retrograde and direct movement.

The rest of the dates recorded are again New Year's Days, except for Trajan 18, where the New Year's Day is significant, and Hadrian 17, where the New Year's Day is omitted.[39]

The list of significant dates in P is given below:

b

	text			calculation	
Aug. year	month	day	sign	position	calculat. −text
16	7	1	♑	♐ 26	− 4
	10	5	♐	♐ 25 ↓	− 5
17	3	30	♑	♐ 26	− 4
19	6	6	♒	♑ 25	− 5
20	1	25	♒	♑ 26 ↕	− 4
21	8	14	♓	♒ 24	− 6
	<12	15	♒ >	♒ 24 ↓	[− 6]
22	5	2[7]	♓	♒ 26	− 4
23	[12]	5	♓	♓ 23 ↕	[− 7]
24	7	4	♈	♓ 23	− 7
26	8	22	♉	♈ 24	− 6
28	10	[3]	♊	♉ 24	− 6
30	[1]1	26	♋	♊ 26	[− 4]
31	6	18	♊	♊ 26 ↓	− 4
	7	12	♋	♊ 26	− 4
33	1	26	♌	♋ 28	− 2
36	[1	1]	♍	♌ 28	[− 2]
38	3	7	♎	♍ 29	− 1
	8	18	♍	♍ 29 ↓	− 1
	11	27	♎	♍ 28	− 2
41	2	30	♏	♏ 0	0

showing the consistent description of the movement of b from sign to sign (only one omission in the year 21 and one stationary point right on the border-line of ♒ and ♓ in the years 20 and 23, respectively). The positions in both S and P deviate in the average by about − 4° from calculation (cf. Pl. 14).

The years in P when b does not leave the interior of a sign are shown at the head of the following column. The signs agree in all cases with the calculations, but it is difficult to say which positions are considered. The most natural assumption would be the middle of a sign, if we remember the analogous case of ♃ (see p. 234). The average, however, is 8½ with oscillations from 3° to 19°. If we add the 4 degrees corresponding to the average deviation of the significant dates, we get only 12°, instead of 15°. One might therefore consider quite different dates as heliacal

b

	text			calculated position	
Aug. year	month	day	sign		
15	3	21	♐	♐	5
18	4	28	♑	♑	7
25	2	8	♈	♈	3 ↓
27	1	15	♉	♉	5 ↓
29	2	13	♊	♊	4 ↓
32	3	24	♋	♋	14 ↓
35	1	9	♌	♌	19
37	1	1	♍	♍	10
39	1	23	♎	♎	3
40	2	5	♎	♎	15

risings or settings but our material is not extensive enough to prove any assumption of this kind. This group of dates in the Saturn movement seems therefore to follow a different principle from that in all other cases.

We might summarize our results in the following form: the "significant" dates follow the same principle for all five planets. In the case of ♃ and b, these dates are supplemented in S by the addition of all New Year's Days. In P only those years are inserted which would otherwise have been omitted, but our material is not sufficient to determine which positions were then considered.

B. HISTORICAL COMMENTARY.

6. Introduction.

The question usually raised about the astrological purpose of ancient astronomical texts seems to me to be without much interest. Our texts do not contain any reference to astrology and the fact that the positions of the planets in the zodiac are of astrological significance does not prove that these texts would not have been written without any interest in astrology. We can say even more: if these texts are of old-Egyptian origin then astrological purposes can certainly be excluded because no astrology existed in Egypt before the latest period of its history.

Much more serious, however, is the question whether our texts actually tell us something about Egyptian astronomy or whether they are merely Hellenistic science in Egyptian disguise; at least one Greek element is indeed obvious, the zodiacal signs.[40]

---

[39] See the apparatus criticus.

[40] It is well known that the division of the ecliptic into twelve parts replaced not earlier than in the latest period the belt of the 36 "decans" of the old-Egyptian star lists; see e.g. Brugsch,

We have three methods of investigating this problem: we can try to trace our text back to old-Egyptian documents; we might compare their methods with the known Greek methods; and finally we can investigate the ancient reports about Egyptian astronomy.

The first method cannot be used because the only available old-Egyptian astronomical sources are of purely mythological character; only the discovery of an hieratic papyrus could bring new light here. The second way would require the knowledge of the method by which Demotic and Greek planetary texts were computed. The little that can be said about this point will be discussed in the last part of this commentary (p. 239). There remains the investigation of the ancient tradition concerning Egyptian astronomy, to which this second part of the commentary will be devoted.

We have two different groups of references to Egyptian science: one rather frequent kind of quotation praises Egypt as the source of deep wisdom in astronomy, mathematics, philosophy and medicine but does not mention a single concrete fact; we therefore do not need to collect these passages here. There exists however a small second group of references, which at least mention the planets explicitly, and which in a few cases even seem to contain detailed information. We proceed to discuss these instances in their chronological order.

7. Aristotle, Diodorus, Pliny.

Aristotle reports in his "Meteorology" [41] (about 350 B.C.) that the Egyptians observed that the planets can come into conjunction both with each other and with fixed stars; he obviously had the same fact in mind when, in "De coelo" II,12,[42] he refers to an occultation of Mars by the moon, observed by himself, adding that the same is reported about the other planets [43] by those from whom "we have many well confirmed notices about every planet," [43] namely, from the Egyptians and Babylonians.[44]

Diodorus (first century B.C.) prefers a more colorful picture.[45] He omits the concrete fact of the mutual occultation, but proceeds thereafter on very much the same line as Aristotle, decorating the report about

the observations of the planets with the trivial remark that the Egyptians observed "movement, periods and stationary points" of the planets [46] and adding that [47] "while they are often successful in predicting to men the events which are going to befall them in the course of their lives, not infrequently they foretell destruction of the crops or, on the other hand, abundant yields, and pestilences that are to attack men or beasts, and as a result of their long observations they have prior knowledge of earthquakes and floods, of the risings of the comets, and of all things which the ordinary man looks upon as beyond all finding out. And according to them the Chaldaeans of Babylon, being colonists from Egypt, enjoy the fame which they have for their astrology because they learned that science from the priests of Egypt." [48] This is more than sufficient to characterize the knowledge of Diodorus both in astronomy and history.

An absolutely isolated statement appears in *Pliny* (first cent. A.D.); he tells us [49] that Petosiris and Nechepso attribute to one degree of the moon's orbit "a little more than 33 stadia," twice as much for Saturn, and the mean value for the sun. The phrase "a little more than 33 stadia" means obviously that 360 degrees are not $33 \cdot 360 = 11880$ stadia long but 12000, and correspondingly the sun's orbit 18000 stadia, the orbit of Saturn 24000. I do not know what is the source of Pliny's report; the reference to Nechepso-Petosiris is of course no proof of Egyptian origin.

8. Theon Smyrnaeus.

A more detailed discussion is needed of a remark of Theon Smyrnaeus (first half second century A.D.), where he speaks [50] about the methods by which the Babylonians and Egyptians succeeded in predicting the celestial phenomena. He says that the Chaldaeans did it by arithmetical methods, the Egyptians by graphical methods, but that both were insufficient as far as the physical explanation is concerned, which the Greek finally gave.[51]

To characterize Babylonian astronomy as "arithmetical" is the best possible description of the astronomical cuneiform texts of Seleucid times; that the Greeks thought of their own cinematical models as physical explanations is easy to understand. One might therefore conclude that Theon was also well informed about Egyptian astronomy. It is, however, very difficult to connect any real sense with his descrip-

Aeg. p. 346 f. or Sethe, Zeitrechnung p. 104. For the present state of the investigation on the history of zodiac see Schott [1].

[41] A 343^b 28 (ed. F. H. Fobes, Cambridge, Mass., Harvard University Press, 1919).

[42] 292^a 9.

[43] "Αστρον means here doubtless "planet" and not merely "star."

[44] As far as the Babylonians are concerned the statement of Aristotle is confirmed by cuneiform texts: cf. e.g. Thompson Rep. No. 241 (occultation of Mars by the moon), No. 192 and 193; or Harper, Letters 565 = Waterman RC I p. 401 (occultation of Jupiter by the moon) and analogous observations quoted in No. 194, 195, 245. Not a single report about an occultation or even about solar or lunar eclipses is preserved from Egypt.

[45] I,81.

[46] τάς τε τῶν πλανήτων ἀστέρων κινήσεις καὶ περιόδους καὶ στηριγμούς.

[47] I quote from Oldfather's translation in the Loeb Classical Library I p. 279.

[48] This last statement is repeated from I,28.

[49] NH II,23 (21) ed. Jan-Mayhoff I, p. 155 f.

[50] Ch. 30 ed. Martin p. 272, ed. Hiller p. 177, transl. Dupuis p. 286 f.

[51] φέροντες οἱ μὲν ἀριθμητικάς, ὥσπερ Χαλδαῖοι, μεθόδους, οἱ δὲ καὶ γραμμικάς, ὥσπερ Αἰγύπτιοι, πάντες μὲν ἄνευ ἀτελεῖς ποιούμενοι τὰς μεθόδους, δέον ἅμα καὶ φυσικῶς περὶ τούτων ἐπισκοπεῖν.

tion of Egyptian methods as "γραμμικῶς," i.e. "graphical" or "geometrical." One can hardly believe that the Egyptians used such descriptive geometrical or nomographical devices as Ptolemy, e.g. in his "Analemma," and even then the application of such graphical methods to the problems of planetary movements would be unintelligible. There remains therefore only the less technical translation of γραμμικῶς as "geometrical," in contrast to arithmetical.[52]

We have thus reduced Theon's statement to an arbitrary historical construction which can be shown to play a decisive rôle in the ancient tradition about the origin of mathematics; the Egyptians are designated fathers of geometry, while the discovery of arithmetic is attributed to Phoenicia. It is obvious in itself that this story is pure fiction even if we did not have both Egyptian and cuneiform texts which make clearly evident the fact that in no approximation can such a clear-cut history of mathematics be justified. On the other hand, the nature of this story points clearly towards Egypt as the place of origin[53] and it seems even possible to localize it still more precisely, namely in Herodotus' famous report on Egypt. There[54] he (or his source) tells us "that the inundation of the Nile created the necessity of re-iterated surveying" and he adds: "it seems to me (δοκέει δέ μοι) that geometry originated in this way and came afterwards to the Greeks." Since then, this explanation of the origin of geometry has been repeated again and again: Diodorus,[55] Strabo[56] (time of Augustus), Heron[57] (first century A.D.[58]), Jamblichus[59] (about 300 A.D.), Servius[60] (about 400 A.D.) and Proclus[61] (fifth century A.D.) are examples. The origin of arithmetic is attributed to the Phoenicians for the first time by Strabo;[56] there followed Porphyrius[62] (third century A.D.), Jamblichus[59] and Proclus.[61] As early as in Diodorus[63] Pythagoras is connected with this Egyptian tradition, then Thales (Heron[64] and Proclus[61]) and Democrit (Diodorus[65] and Diogenes

Laertius[66] third century A.D.), a fact which shows still more the unreliability of these historical fictions.

Theon's remark appears therefore to be no more than the extension, without any concrete background, of a generally accepted assumption about the origin of the different parts of mathematics to astronomy.

9. Clemens Alexandrinus.

The situation in the case of Clemens Alexandrinus (second half of the second century A.D.) is very different. Here again we meet the famous "Hermetic" corpus, when he describes,[67] doubtless as an eyewitness, the procession of the Egyptian priests. He tells us that a priest, called ὡροσκόπος, marches in the procession wearing plumb-line and sighting instrument as symbols of astronomy,[68] and he adds that this priest is supposed to know by heart the four hermetic books (τὰ Ἑρμοῦ βίβλιου τέσσερα). The titles of these four books are: on the arrangement (διακόσμος) of the fixed stars; on the position (τάξις) of the sun, moon and five planets;[69] on the syzygies (σύνοδοι) and phases (φωτισμοί) of the sun and moon and, finally, on the risings (ἀνατολαί). In the following, additional "hermetic" books are mentioned, dealing with geography, administration of the temples and related topics.

We must now try to decide whether the books mentioned really belong to the comparatively recent group of literature connected with the name of "Hermes trismegistos"[70] or whether they can be traced back to original Egyptian sources. The fact that Clemens himself uses the name of "Hermes" does not prove

---

[52] Cf. e.g. Jamblichus, Vita Pyth. 29, 158 (ed. Deubner p. 89): "It seems that τὰ περὶ τὰς γραμμὰς θεωρήματα [the theorems concerning the lines] originated with the Egyptians; logistics and numbers, however, with the Phoenicians."

[53] There is no doubt that Egypt was in general much more familiar to the Greeks than Babylon. The fact that the reports of Herodotus and Strabo about surveying in Egypt correspond to real observations was emphasized by Lyons [1].

[54] II,109.

[55] I,81 (cf. I,94).

[56] XVI,2,24 and XVII,1,3.

[57] Opera IV, ed. Heiberg p. 176, p. 398.

[58] Concerning this date see Neugebauer [3] p. 21 ff.

[59] Vita Pythag. 29, 158 (ed. Deubner p. 89).

[60] In Verg. Buc. III,41 ed. Thilo-Hagen III,2 p. 57. The vulgata-redaction gives an even clearer text; see Wallace [1] p. 26.

[61] In Eucl., ed. Friedlein p. 64 f.

[62] Vita Pythag., ed. Nauck p. 19.

[63] I,69,4; I,96,2; I,98,2 and X,6,4; cf. I,94,3.

[64] Definitiones. Heron opera IV ed. Heiberg p. 108.

[65] I,96,2 and I,98,3.

[66] IX,35 (Diels VS 68 [55] A 1).

[67] Strom. VI,4 (ed. Stählin I p. 448 f., transl. Overbeck p. 315 f.).

[68] Ὡρολόγιόν τε μετὰ χεῖρα καὶ φοίνικα ἀστρολογίας ἔχων σύμβολα Borchardt recognized that ὡρολόγιον and φοῖνιξ are plumb-line and sighting instrument, copies of which are preserved in the Berlin museum (Borchardt [2] p. 12 and [1] p. 54 and pl. 16; reproduced in Sloley [1] pl. XVI). The Egyptian name of the first instrument is, according to Borchardt, mrḫ.t "instrument to measure" (so according to Sethe, quoted by Borchardt [1] p. 54 note 1) or "instrument to know (viz. the hour)" if we derive mrḫ.t from rḫ "to know" instead of from rḫ "number." The Egyptian name for the φοίνιξ ἀστρολογίας was discovered by Spiegelberg ([4] p. 113 f.) to be b'j n imj wnw.t "the palm rib of the man concerned with the time," which shows the accuracy of the Greek translations and the correctness of Clemens' report. As far as the ὡρολόγιον is concerned another interpretation is perhaps possible, namely "waterclock." This meaning of the word is testified by Pap. Oxy. III,470 (Borchardt [1] p. 10 ff.) and offerings of this instrument are represented on different wall paintings (Pogo [1] Figs. 5 and 6). The mrḫ.t may not have been mentioned explicitly, being merely an auxiliary part of the φοῖνιξ.

[69] I follow here the arrangement of the text given by Stählin. Overbeck translates the same passage twice: first at its place in the manuscripts and then at the place required by Stählin's emendation!

[70] A large part of the extant remains of this literature is collected in the great work of W. Scott "Hermetica." Cf. furthermore: RE 8, 792 ff.; Gundel, Hermes trismegistos and Cumont [1]. Gundel traces the star catalogue of his text back to the third century B.C. (Gundel, Hermes trismegistos p. 131 ff.).

anything, because at his time "hermetic books" meant about the same as "Egyptian theological literature." There are indeed different arguments in favor of the assumption that the holy books worn in the procession are not of Hellenistic origin as the hermetic corpus is. The main argument is, of course, the general historical consideration; it would be very unlikely to assume that the Egyptian priests replaced their own old sacred texts by very recent ones. The second argument is based on the agreement of the booklist given by Clemens with a hieroglyphic list of "books" from the "library-room" in the Horus temple of Edfu, built by Euergetes II (145 to 116 B.C.[71]). Although even the Edfu-list does not take us beyond the limits of Hellenism we shall indicate relations between this text and inscriptions of the New Kingdom. We have therefore first of all to establish the close agreement between Clemens' list and the Edfu catalogue.[72] This is shown by the following comparison:

I think the agreement between Clemens and the Edfu list is close enough to justify the statement that the "books" mentioned by Clemens belong to essentially the same type of documents as the collection catalogued in Edfu. The next step must therefore be to get some more information about the character of the Edfu texts.

This is possible, although to a very modest extent, from the appearance of one of the Edfu booktitles in the Demotic papyrus Carlsberg no. 1,[79] namely "on the protection of the bedroom." [80] This papyrus, although written in Roman times, contains commentaries on the inscriptions around the picture of the sky-goddess Nut, found in the cenotaph of Seti I (ca. 1300 B.C.) and in the tomb of Ramesses IV (ca. 1160 B.C.). It is very unlikely that the "books," quoted as authorities for the explanations given in the papyrus, should be 1000 years younger than the text to which they belong. It is therefore practically certain that the books mentioned by Clemens and

Edfu-"library" [73]	Clemens [74]
—	ὑμνοι θεῶν (448, 28)
the protection of the king in his house (7)	ἐκλογισμὸν τοῦ βασιλικοῦ βίου (448, 28 f.)
rule of the (periodic) repetition [75] of the stars (10)	περὶ τοῦ διακόσμου τῶν ἀπλανῶν φαινομένων ἄστρων (449, 2f.)
to know the (periodic) repetition [75] of the two lights (= sun and moon) (9) [76]	περὶ τῆς τάξεως τοῦ ἡλίου καὶ τῆς σελήνης καὶ περὶ τῶν πέντε πλανωμένων (449, 3f.) περὶ τῶν συνόδων καὶ φωτισμῶν ἡλίου καὶ σιλήνης (449, 4f.)
—	περὶ τῶν ἀνατολῶν (449, 6)
all districts [77] and to know their content (11)	περὶ τῆς κοσμογραφίας καὶ γεωγραφίας (καὶ) χωρογραφίας τῆς Αἰγύπτου καὶ τῆς τοῦ Νείλου διαγραφῆς (449, 10ff.)
on the administration [78] of the temple (4) and the temple personnel (5)	περὶ τῆς κατασκευῆς τῶν ἱερῶν καὶ τῶν ἀφιερωμένων αὐτοῖς χωρίων (449, 13)
every protection of the procession of your majesty outside of your temple on your festivals (12)	περὶ θυμάτων ἀπαρχῶν, ὑμνων, εὐχῶν, πομπῶν, ἑορτῶν καὶ τῶν τούτοις ὁμοίων (449, 19f.)

[71] Cf. A. Bouché-Leclercq, Histoire des Lagides II p. 84 f.
[72] Already recognized by Brugsch, Aeg. p. 157.
[73] Published by Bergmann HI Pl. 64 f. and Chassinat [1] p. 347 f. and p. 351. Translations: Bergmann HI p. 46 ff. and Brugsch Aeg. p. 156 f. The numbers in ( ) refer to the arrangement in Brugsch Aeg. p. 156 f. Dr. H. Ranke was so kind as to draw my attention to Chassinat's publication.
[74] The numbers in ( ) refer to page and line in the edition of Stählin.
[75] Egyptian: whm.

[76] Brugsch erroneously (8).
[77] Bergmann translates "Die Aufzählung aller Orte . . ." and Brugsch follows him, but there is no word corresponding to "Aufzählung" in the text.
[78] Egyptian: śśm.
[79] Nr. 26 in Brugsch's enumeration (Bergmann HI Pl. 46,3) and Lange-Neugebauer [1] p. 12.
[80] Egyptian: s3 ḥnk.t. The "bedroom" is also a room in the temple, especially where the bier of Osiris stands (cf. WB III, 119).

those in the Edfu temple are of purely Egyptian origin, since they have about the same character as the texts quoted in the Pap. Carlsberg 1. But this admitted, we are able to form at least a vague idea about their content from the quotations in Pap. Carlsberg 1. In all instances, namely, where the quotations in the papyrus give us some glimpse of the content, its character appears to be purely mythological. "Book" probably meant even such comparatively short rolls of papyrus as Carlsberg 1, containing mythological formulations of very simple astronomical facts. The title of the book "on the 5 epagomenae" mentioned in Pap. Carlsberg 1 [81] therefore does not prove anything at all of a serious astronomical content.

We thus come to the following conclusion: the books mentioned by Clemens are very likely of old-Egyptian origin, related to known texts from the tombs of the New Kingdom, but for this very reason only concerned with theoretical astronomy, in spite of some titles which at a first glance might point in this direction. In other words, our best informed source from Greek literature gives us exactly as little information about Egyptian theoretical astronomy as the available Egyptian texts themselves.

10. Macrobius.

The last author we should mention is Macrobius (ca. 400 A.D.). The date of this author alone makes it very unlikely that he might reveal any essential new material; but the fact which shakes our confidence in this source most of all is the long and minute report about the Egyptian invention for subdividing the orbit of the sun into twelve parts.[82] Here every word is idle construction absolutely contrary to the true history of the zodiac.

The planetary theory which Macrobius ascribes to the Egyptians is well-known to be the following [83]: Mercury and Venus are satellites of the sun, but the moon, sun and the three other planets revolve around the earth. I do not need to discuss this theory in more detail, because this is done very carefully by Heath,[84] Dreyer [85] and others. The conclusion reached is that this model probably is the invention of Heraclides Ponticus (4th cent. A.D.) and at any rate arbitrarily credited to the Egyptians by Macrobius.

C. The "Eternal Tables"

11. The result of the investigations in the preceding section is that we can not gain any information about Egyptian astronomy from the classical literary

sources.[86] However, we owe to Ptolemy a short report on earlier planetary theories which might have some bearing on our present problem.[87] The second chapter of book IX of the Almagest is devoted to introductory remarks about the fundamental difficulties in the theory of planets and earlier attempts to overcome them. Ptolemy tells us first that the only consistent sequences of observations concern "the stationary points and heliacal risings and settings." This refers undoubtedly to the Babylonian planetary tablets which are built precisely in this principle. Then he goes on to speak about Hipparchus, who wisely restricted himself to systematic collection and arrangement of observations "thus proving that the phenomena do not agree with the assumptions of the astronomers of his time." Ptolemy mentions thereafter that these astronomers made such rough simplifications of the problem as to assume constant anomalies and constant amount of retrograde orbits in order to construct their excentric or epicyclic models, assumptions which could not satisfy a scholar like Hipparchus; but the assumptions of practically all those who tried to demonstrate the existence of uniform movements on circles by using the so-called "eternal tables" were of this kind.[88]

Ptolemy's contempt for these αἰώνιαι κανονοποιίαι is obvious. He does not reveal more about their character but it seems to me that it follows from the passage quoted that these tablets have been composed by a method other than the assumption of circular movements since he states they were used in order to derive *from them* (διά) proofs of the existence of such movements. It seems possible to me that the planetary tables, discussed here, may belong to this elementary type of "eternal tables." The arguments which speak in favor of this assumption are of different character. The first is direct literary evidence from a Greek horoscope,[89] the date of birth being 81 A.D., where we read in the introduction that the Egyptians had recorded the movements of the seven planets (lit. "of the seven gods") and have "handed us down their knowledge about them without envy by means of the eternal tables." Secondly, we have a Greek text, also from the time immediately before Ptolemy, which is almost a complete copy of a part

---

[81] II,12. Lange-Neugebauer [1] p. 23.
[82] Somnium Sci. I,21, 12 ff.
[83] Somnium Sci. I,19.
[84] Heath, Aristarch p. 259.
[85] Dreyer [1] p. 129.

[86] This fact is unintentionally illustrated by the absurdities accepted as "L'astronomie égyptienne" in the book of E.-M. Antoniadi (Paris, Gauthier Villars 1934) of the same title, relying exclusively on ancient fanciful stories and ignoring the last hundred years of research in Egyptology.
[87] One of the most important sources on Hellenistic-Oriental astronomy, Vettius Valens (second century A.D.) is practically unexamined, as far as information about exact astronomy is concerned. This problem very much deserves serious consideration but goes far beyond the limits of the present edition.
[88] Ptolemaeus, Almagest IX,2 (ed. Heiberg I,2 p. 211, 4 ff.): τούτοις γὰρ ἐπιβεβλήκασι μὲν σχεδόν, ὅσοι διὰ τῆς καλουμένης αἰωνίου κανονοποιίας τὴν ὁμαλὴν καὶ ἐγκύκλιον κίνησιν ἠθέλησαν ἐνδείξασθαι.
[89] Published by Kenyon, Cat. Pap. Br. Mus. I, No. 130, Col. I, p. 133.

of the Stobart tables, showing us that we are right in referring Ptolemy's statement to Greek as well as to Demotic planetary tables. The third method of approach is the investigation of the means of obtaining such planetary tables; we begin our discussion with this last problem.

The main difficulty in attempting to secure information regarding the methods used to compose the planetary texts lies in the fact that text and modern calculation are in general in good agreement. The truth of this statement, which perhaps sounds paradoxical, will be evident from the following remark. Suppose the positions given by a text coincide exactly with the true positions. Such a faultless list of positions would only show that the methods used (either theoretical or instrumental) are perfect, but could never reveal anything about the special character of the method. Only some kind of systematic deviation of the text from the facts can give us hints as to the type of underlying procedure. Now the only systematic deviation which can be recognized in our texts is the practically constant amount by which the longitudes given in the text are in advance of calculated longitudes, which can be accounted for by the assumption that a slightly different fixed vernal-point has been used.

There is only one point which follows with certainty from the texts: the entrance of a planet into a sign is indicated even if it is invisible because of the near-by positions of the sun. Therefore some kind of "theory" must have been used. It seems to me, however, a premature conclusion to assume from this fact that all positions must have been calculated.[90] I now must add that the agreement between text and calculation is good only in the main parts of the orbits (the direct or "linear" part, if we look at the diagrams on Pls. 1 to 15). In the retrograde parts, however, we found considerable discrepancies.[91] This is not strange in itself, but it is very difficult to explain how one consistent theory could provide on the one hand only a rough description of the retrograde movement, but should give on the other hand very satisfactory results for all the remaining parts, and without any systematic deviation from the truth even after many years. Such a simple theory as the moon-theory in Pap. Carlsberg 9, based on a single period-relation, is out of the question in the case of planetary movement, without resulting in large systematic (yet periodic) deviations the non-existence of which is shown by our calculations. On the other hand a highly developed theory such as the Babylonian or Ptolemy's planetary-theory would furnish better results in all parts of the orbit if it worked as well in the largest part as we have already seen. The sim-

---

[90] Wislicenus held it highly probable "dass die ganze Tafel (i.e. P) berechnet und nicht beobachtet ist" (Spiegelberg DPB p. 30). Biot (in Brugsch [1] p. 63 f.) also believes that S is calculated because "il faudrait pour cela (i.e. observations) qu'au temps de Trajan il eut existé à Thèbes ou à Memphis, un grand observatoire fixe, desservi par des observateurs attitrés, munis d'instruments . . .". These arguments are obviously absurd because one single man can easily recognize the days when a planet enters a new sign and the only instrument required is some kind of simple astrolabe.

[91] Cf. p. 231.

V,2–4	♎ :	GHG restore erroneously αιγο instead of ζυγο.
VI,13	4 :	S (C₁ obv. IV,13) has 2 instead of 4 (GHG say p. 26 erroneously "Venus" instead of Mercury and "14th year" instead of 10th).
VI,13	[20 :	GHG restore 18 according to S but preceding and following numbers in T require 20 and not 18.
VI,15	9 :	S (C₁ obv. V,13) has 5 instead of 9 (GHG say p. 28 erroneously "Venus" instead of Mercury and "15th year" instead of 11th).
VI,15	2[0 :	S (C₁ obv. V,14) has 15 instead of 20 (GHG say p. 26 erroneously 18 instead of 15).
VI,25	21 :	S (C₁ obv. V,25) has 20.
VI,26	[8] :	GHG restore here [3] because of an error of Brugsch.
VI,26	♍ :	GHG p. 26: "The meaning of the astronomical sign after παρ is not clear. It resembles a badly written θ but there is a space between it and παρ."
II′,4	23 :	S (C₁ rev. III,24) has 24.
II′,23	IX 4 :	The remainders in S, C₁ rev. V,10 seem to fit only to the date VII,4 and not IX,4. Calculation shows that IX,4 (⊁ 22) is better than VII,4 (⊁ 15).
III′,5	[5 :	Restoration according to S. GHG p. 27 say "We hesitate to restore the figure since the extant entries . . . do not agree precisely with those in S."

III′,9	[1]2 :	S (C₁ rev. III,28) has 15.
III′,11	30 :	S (C₁ rev. III,29) has 26.
III′,11	♉ :	GHG p. 27: "The astronomical sign following αιγο perhaps denotes that the planet had gone backward instead of forward."
III′,18	[1]7 :	S (C₁ rev. V,13) has 18.
III′,21	♒ :	GHG erroneously 6 ⊁ (cf. following line).
III′,22	6 ⊁ :	GHG erroneously ⊁ only (cf. preceding line). S (C₁ rev. V,14) has 5 (the corresponding month is destroyed in S) calculation gives ≈ 25.
IV′,7	[25 :	Brugsch hesitated between 21 and 25. GHG restored 21 but 25 is required by calculation (♐ 28).
IV′,10	13 :	S (C₁ rev. IV,7) has 1[2] (only 1[2] or 1[7] are possible readings of the remaining traces in S).
IV′,12	7 :	S (C₁ rev. IV,8) has 6.
I″,1	[(year) 18 :	This date proposed by GHG p. 26 f. is practically certain although the entries of T are two days later than those of S (C₂ obv. V,24 to 28). Calculation gives:

$$
\begin{array}{llll}
\text{S} & \text{VI} & 17 & ♉ \ 28 \\
\text{T} & & 19 & ≈ \ 2 \\
\text{S} & \text{VII} & 5 & ≈ \ 28 \\
\text{T} & & 7 & ⊁ \ 2 \\
\text{S} & & 21 & ♈ \ 0.
\end{array}
$$

Pap. Tebtunis II, 274.

[I] (month)	[II to IV] (♄ ♃ ♂)	V (♀)	VI (☿)	VII (☾)
[(year) 10 (or) 29]				
[I]	destroyed	[♎]	[11 ♎]	destroyed
[II]		[18 ♍ 23 ♎]	♎	
[III]		[♎]	17 [♏]	
[IV]		[11 ♏]	4 ♐ 23 [♑]	
[V]		[11] ♐	11 ♒	
[VI]		[6 ♑]	♒	
[VII]		[1 ♒ 25] ♓	14 ♓	
[VIII]		[19 ♈]	1 ♈ 17 [♉]	
[IX]		[14 ♉]	6 ♊	
[X]		[9 ♊]	♊	
[XI]		[3 ♋ 27 ♌]	[14] ♋	
[XII]		[21] ♍	4 ♌ [20 ♍]	
[(year) 11 (or) 30]				
[I]	destroyed	[9 ♎]	9 ♎ 2[0 ♍]	destroyed
[II]		[4 ♏ 28] ♐	16 ♎	
[III]		[23 ♑]	8 ♏ [27 ♐]	
[IV]		[18 ♒]	[15 ♑]	
[V]		[13 ♓]	[♑]	
[VI]		[8 ♈]	[20 ♒]	//////////////
[VII]		[11 ♉]	[3 ♓ 23] ♈	
[VIII]		[♉]	[♈]	4 ////
[IX]		[♉]	[7] ♉	
[X]		[♉]	[12] ♊	3 6
[XI]		[14 ♊]	[5 ♋] 21 ♌	
[XII]		[14 ♋]	[8] ♍ θ	2 5

fragm. (a) and (b)

I′ (month)	II′ (♄)	III′ (♃)	IV′ (♂)
[(year) 14 (or) 33]			
I	[♒]	[♐]	[♍]
II	[♒]	[♐]	[13 ♎]
III	23 [♊]	[♐]	[♎]
IV	♓	[5 ♑]	[1 ♏]
V	♓	[♑]	[13 ♐]
VI	♓	[♑]	[25 ♑]
VII	♓	♑	[♑]
VIII	♓	[1]2 ♒	3 [♒]
IX	♓	♒	13 [♓]
X	♓	30 ♑ς	[♓]
XI	♓	♑	7 [♈]
XII	♓	♑	[♈]
(year) 15 (or) 34			
I	♓	♑	[♈]
II	♓	♑	[♈]
III	♓	♑	[♈]
IV	♓	[1]7 ♒	[♈]
V	♓	♒	2[5 ♉]
VI	♓	♒	[♉]
VII	♓	♒	[11 ♊]
VIII	♓	6 ♓	[27 ♋]
IX	4 ♈	♓	[♋]
X	♈	♓	[21 ♌]
XI	♈	♓	[♌]
XII	♈	[♓]	[15 ♍]

fragm. (d)

I″ (month)	VI″ (☿)
[(year) 18 (or) 37]	
////	//////////////////////
[V]	//// ♑
[VI]	19 ♒
[VII]	7 ♓ . [//// ♈]
[VIII]	//// ♉
////	//////////////////////

fragm. (c)

plest assumption seems to me therefore the hypothesis that our texts are the result of combined observations and calculations, which would neither require any elaborate theory about the structure of the planetary system nor any observation technique, but would explain why systematic deviations from the facts never occur. On the other hand, as Ptolemy says, such lists of consecutive positions could be used as basic material for establishing a cinematic model.

To consider Demotic texts as some of the so-called "eternal tables" in Ptolemy's report would have to meet the serious objection of the difference in language unless we did not have the explicit proof of texts of the same kind written in Greek. Such a text was actually discovered among the Tebtunis papyri, now in possession of the University of California, published under the cooperation of T. G. Smyly by Greenfell, Hunt and Goodspeed in 1907 in the second part of the "Tebtunis Papyri" as no. 274. As the editors recognized, these papyrus-fragments contain planetary positions almost identical with positions given in the Stobart tablet C_1, belonging to the years 10 to 18 of Trajan (107 to 115 A.D.). In order to make this papyrus readily available for further discussions I give here its complete transliteration and comments, including those which are already contained in the edition quoted above.

12. Papyrus Tebtunis II,274.

The numeration of the columns adopted, p. 241, is different from the one in the edition. For the ruling of the original see plate 25. On fragment "a" a small part on the left hand lower corner is now missing.

The years heading the single sections are in the first place the regnal years of Trajan, and secondly years since Titus 1 (79 A.D.), an otherwise unknown era. There is only one additional instance where this era appears to be used, namely in Pap. Brit. Mus. No. 130,[92] the very same horoscope, mentioned above, which quotes the Egyptian "eternal tables." This horoscope, cast for 81 A.D. but undoubtedly written later, belongs, therefore, to the same period as the Pap. Tebtunis II, 274 (in the following referred to as "T") and the practice of both texts in counting years as regnal years of Titus might even indicate a direct connection between the two texts.

The very close agreement between the dates in T and the Stobart tables shows that the calendar em-

ployed is the Alexandrian calendar. Although the arrangement in T is a little different from S, the meaning of the dates is the same. A passage as in VI line 5 ff.

	V	4	$\nearrow$	23	$\eth$
	VI	11	$\approx$		
	VII		$\approx$		

means that the planet (here Mercury) enters on the 4th of month V the sign $\nearrow$, on the 23rd the sign $\eth$, on the 11th of the next month $\approx$, but stays inside $\approx$ during the month VII.

Comparisons between the positions given by T and S show a slight tendency of S to place the entrance of a planet into a sign earlier than T, but this might be purely accidental in our little fragments (in 10 instances it is earlier, averaging one day; in 3 instances later; in 12 there is exact agreement). As far as the agreement with modern calculation is concerned neither S nor T shows any essential difference.

13. T goes beyond S by containing a column referring to the moon. The assumption of the editors that the dates

year 11	VIII	4	////
	X	3	6
	XII	2	5

give the dates of the actual new moon and the subsequent visible crescent can easily be proved correct by calculation. These new moon dates, missing in the planetary texts P and S, enable us to establish an interesting relationship with another Demotic manuscript namely the Papyrus Carlsberg 9. This text,[93] written after 144 A.D., but undoubtedly of much older origin, gives a simple method for determining the new moon dates by a cyclic calculation of exactly 25 Egyptian years as period. The dates given by this papyrus show the same feature as the dates in T, namely decreasing by 1 every second month, leaving the dates in the intermediate months undetermined. This resemblance between T and the Pap. Carlsberg 9 (referred to in the following as "C") invites us to investigate the dates in detail.

The dates in C being expressed in the Egyptian calendar, we first have to convert the Alexandrian dates in T into Egyptian dates. This gives the correspondence:

Trajan 11	Alex. cal.	VIII	4	Jul. cal.	March 30	Eg. cal.	IX	7
= 108 A.D.		X	3		May 28		XI	6
		XII	2		July 26		epag.	5

The 25-year cycles given in C begin with the Egyptian New Year in the years (A.D.) 19, [44], 69, 94, 119,

and 144 such that Trajan 11 belongs to the cycle from 94 (= Domitian 14) to 119 (= Hadrian 3). The

[92] Kenyon, Cat. Pap. Brit. Mus. I p. 133.

[93] Neugebauer-Volten [1].

Egyptian New Year's day in 94 A.D., the first year of the cycle, is July 31 which shows that our dates in 108 A.D. belong to the cycle year 14. If we now look at the Egyptian dates given in C for cycle year 14, we find there [94]

VIII 7   X 6   XII 5

or dates which are exactly one month in advance of the dates in T. This is of course no contradiction, but merely says that the scheme in T considers those dates as determined which C leaves undetermined and vice versa; but the parallelism in method is obvious.

14. The preceding remarks are of importance for our problem because we now know the calculation of at least one column in T, since the structure of C is entirely clear, being based exclusively on the single period relation

25 Egyptian years = 309 synodic months

which does not involve any theoretical considerations about the movement of the moon. It is very unlikely that the columns referring to the planets are calculated according to some elaborate theory if the moon is merely treated in such a simple way. In other words, the relationship between T and C on the one hand and between T and S on the other is a very strong argument for the assumption that the planetary texts were computed by methods independent of cinematic models using excenters and epicycles. This is exactly the same result we arrived at by investigating the deviations of the recorded positions from the positions calculated by modern means. All this would fit into the picture one would expect from the remarks about the "eternal tables."

D. CONCLUSIONS.

15. If we are right in our assumption that all our planetary texts belong to the class of the so-called "eternal tables," their date of origin would be pushed back at least to before 200 B.C., because Apollonius of Perga had already used the method of excenters and epicycles.[95] This date ante quem is furthermore supported by the following consideration. We showed it to be very probable that our texts used a fixed division of the ecliptic, corresponding to the longitude of about $\lambda = -4$ in the time of Augustus (cf. p. 231). At the time of Hipparchus this vernal-point would therefore have the longitude of about $\lambda = -5$. It is now possible, on the other hand, to determine the position of Hipparchus' vernal point. In a brilliant paper, H. Vogt succeeded in calculating longitudes and latitudes for 122 stars according to Hipparchus'

lost star catalogue. He proves [96] that Hipparchus' longitudes show no systematic error when referred to the vernal point of -138.[97] Hence, if we do not want to make the ad hoc assumption of some otherwise unknown Greek system, which deviates by five degrees from that of Hipparchus, this precludes the possibility of deriving the longitudes of the planetary texts from Greek astronomy of the period between Hipparchus and Ptolemy.[98]

16. On the other hand, I do not see any means of determining a still earlier limit for the origin of our planetary tables. However, the first century of the Hellenistic age could have been a period of creation of new astronomical methods in Egypt just as it was the time of origin of the Babylonian theoretical astronomy. And to the very same period points uniformly the now known material of Demotic astrological texts if we try to determine the time of introduction of astrology into Egypt.[99] This does not of course preclude a very strong Egyptian component as the astrological literature shows.[100] On the other hand, both Egyptian and Greek sources make it very unlikely that any serious Egyptian astronomy existed before Hellenistic times, as we have seen in the preceding sections.

As a result of these considerations, the most probable assumption seems to me to be that the planetary tables, discussed here, originated during the first period of the Hellenistic age. The problem of estimating the Greek or Egyptian influence is, as far as I can see, beyond our present limits of knowledge.

§ 5. Terminology, Paleography and Related Topics

A comparative list of the characteristic sign-forms in the planetary text and in some other astronomical papyri and ostraca will be published separately in order to make this palaeographic material available for Demotic studies in general. The intention of the following remarks is merely to emphasize some peculiarities of P and S.

[94] Neugebauer-Volten [1] p. 395.

[95] Ptolemy, Almagest XII,1 (ed. Heiberg I,2 p. 450,10 = Apollonius ed. Heiberg II p. 137 f.).

[96] Vogt [1] col. 25.

[97] Ptolemy's vernal point has the length of about -1; $15°$ with respect to the true equinox of his time (Vogt [1] col. 25).

[98] The "Hermetic" texts, discovered by Gundel a few years ago, support this argument. The longitudes given in this text are smaller than Ptolemy's longitudes, not greater as in our text. The description of the star configurations follows, not Ptolemy, but Hipparchus and even older sources (Gundel, Hermes Trismegistos p. 127). From the fact that the longitudes are smaller by about $2;20°$ to $3;40°$ than those given by Ptolemy (Gundel, Hermes Trismegistos p. 131 and 148 ff.), it therefore follows not only that these longitudes are measured with respect to Hipparchus' coordinate-system, but that even older longitudes refer to an origin corresponding more or less to the vernal-point of their time.

[99] This will be shown in a forthcoming paper.

[100] Cf. Cumont [1] p. 71 ff. (and p. 144 1). Investigating the astrological literature (written in Greek!) he reaches the following conclusion: "les écrits astrologiques les plus anciens témoignent d'une ignorance presque totale de la vie intérieure de cités helléniques."

a. Years.—The year-numbers in P run from 15 to 41 (only 34 and 36 destroyed) and in S from 1 to 19 (7 and 8 destroyed, but 9 to 17 appear twice), so that we have a fairly complete sequence of year-numbers. The sign for "year" seems to be the same in both texts, namely $ḥ\mathit{3.t}\ sp.$[101]

Spiegelberg mentions the following rule [102]: the number signs from 1 to 9, except 6, should have the feminine -t when following a feminine word, but not the numbers beyond 9. Both texts contradict this rule. P adds the -t very irregularly, namely to 20, 29, 38 and 39 (and perhaps to 30 [103]). S, on the contrary, writes -t consistently, even following 6 and beyond 9.[104] Numbers ≥ 10 have the -t just below the 10-sign, which is written in the upper half of the line so that 10-t forms one group of the same height as the following unit.[105]

I was unable to decipher the year-name in S C_2 rev. II,8, which indicates the beginning of the reign of Hadrian. The $ḥ\mathit{3.t}\ sp$ at the beginning is clear, probably followed by 1-t. Brugsch read the remaining signs as "$A - Pa$" [106] which he translated as "grand maison," i.e. "pharao." But before the undoubtedly correct aleph-sign stands another sign which might be the cartouche; furthermore Brugsch's reading "pa" is incorrect and the group above the end is completely disregarded. In spite of the fact, however, that the aleph fits the expected name "Hadrianus" I am not able to explain the remaining signs.

b. Months.—Both texts indicate the months by ordinary number signs as far as month 2 to 12 is concerned. However, S often uses for the first month a special sign which corresponds to the hieroglyphic tpj "first." These cases are marked in our translation of S by italic type 1 instead of 1. P uses ordinary 1 consistently.

The closer investigation of S shows that tablets A and E use tpj consistently (with only one exception in E rev. IV,27). C_1 and C_2 on the contrary have frequently ordinary 1 (in 22 instances 1, in 47 cases tpj), but without any visible rule. The following day-numbers obviously do not have any influence because practically all possible combinations occur in both cases.[107] If however tpj appears then the sign $\odot$ "day" always follows before the date,[108] which never occurs after ordinary 1.

If two consecutive dates belong to the same month, then the number of the month is never repeated in S but always repeated in P.

c. The Epagomenal Days.—As is well known, the Egyptian calendar contains, besides the twelve months of 30 days each, five additional days, in Greek αἱ ἐπαγόμεναι (with or without ἡμέραι). The usual Egyptian name is 5 $ḥrjw\ rnp.t$ "the five (days) on the year," [109] an expression of which the Greek name is the exact translation. Our planetary texts use the name $ḥb$ "festival," as already established by Brugsch,[110] meaning of course "festival of the epagomenae."

Among the 16 instances of the epagomenae in the planetary texts every day number from 1 to 5 occurs but never 6 (which might occur in S using the Alexandrian calendar). In S $ḥb$ is always followed by the sign $\odot$ before the day-number, but never in P.

d. Days.—The most evident palaeographical difference between P and S lies in the writing of the day-numbers. P uses for the ordinary number signs all dates from 1 to 30; in S on the contrary there appears in several instances the sign $\odot$ (to be read as ss) "day" which is always followed by the special day-number-signs, developed, as is well known, from the horizontal number-signs.[111] One case has just been mentioned, namely the epagomenae; we will now investigate the other occasions in S where day-dates are not given by ordinary numbers.

Besides the epagomenae there exists one more simple instance, namely the thirtieth day, which is consistently written as $ss\ 'rḳ$ "last day." Only for the sake of simplicity do I transliterate this sentence by "$\odot$ 30."

We can now direct our attention only to those cases where $\odot$ is used together with dates from 1 to 29 belonging to the twelve ordinary months. Two rules are obviously followed:

Rule one. If two subsequent dates occur in the same month, then the sign $\odot$ replaces the number of the month in the second line. Example (C_2 rev. IV,8 to 10):

9	3	♈
	$\odot$ 27	♉
10	22	♊

This rule is followed in 149 cases and overlooked in only 6 cases.

Rule two. The sign $\odot$ is used in the first line of each group of numbers referring to the same planet. Example (C_2 rev. IV,2 to 4):

	♀	
1	$\odot$ 2	♏
2	10	♐

---

[101] Cf. Gardiner, Gr. p. 204.

[102] Spiegelberg Gr. § 82.

[103] In the case of "year 28" the 8 and what follows is destroyed.

[104] The -t following 9 is placed by P behind the 9, by S below it.

[105] In C_2 rev. I,1 is 19 written as 10-t 9-t.

[106] Brugsch [1] p. 20 and p. 29.

[107] Even 1 1.

[108] See below section d.

[109] WB II,430.

[110] Brugsch [1] p. 21 f. Cf. also Möller, Pal. III No. 512. Spiegelberg overlooked Brugsch's explanation when he wrote (Spiegelberg DPB p. 30 note 2) that "die Lesung der Gruppe nicht feststeht."

[111] Only one and nine are not different from the ordinary signs. Cf. Gardiner Gr. § 259 and WB IV,58 sub verbo sw.

This rule is not as strictly observed as the first one, but is still obvious enough; among 129 cases only 23 exceptions exist. Anyone who knows the carelessness of Egyptian scribes will not be surprised by this rate of omission.[112]

In almost all other instances the sign $\odot$ is not used and consequently neither the special date-sign. But there still remain 49 instances (among about 500) which need explanation, e.g. in a passage like:[113]

$$\begin{vmatrix} 3 & 17 & ⌐ \\ 5 & \odot\ 20 & ♉ \\ 6 & 11 & ≈ \end{vmatrix},$$

where no obvious reason can be seen for putting $\odot$ in the fifth month and nowhere else. What is the reason for this striking inconsistency? In order to answer this question we have to consider the following statistics of all cases of this kind:

month [114]	number of cases of	
	dates < 10	dates ≧ 10[115]
2	3	0
3	3	0
4	1	0
5	2	2
6	2	0
7	0	0
8	1	0
9	0	2
10	23	0
11	2	0
12	0	0

Obviously "month 10" plays an exceptional rôle, a fact which immediately gives the solution of the problem: supposing that 10 were followed by a number n less than ten, then the group could be read as "10 $+\ n$"; in order to avoid such mistakes the sign $\odot$ "day" has been inserted.[116]

The correctness of this explanation can be proved by the following inverse consideration: there occur in S altogether 55 instances of "month 10," among which there are 23 cases where the following day-number is

less than 10. The above list shows that there does not exist a single case where $\odot$ was omitted between 10 and a following unit.

The 4 cases where dates ≧ 10 are preceded by the sign $\odot$ consist exclusively in dates 10 and 20, which shows that here the rule "in order to avoid misreadings never combine a ten-sign directly with units" was followed without actual need. There remain however 14 cases where units are combined with units; but 13 of these cases can be explained simply by the fact that they are preceded or followed (or both) by numbers where $\odot$ has been used according to the preceding rules. Only one case remains as an isolated error.[117]

As mentioned at the beginning, P never uses the special day-numbersigns. There exists however in P another peculiarity. Spiegelberg has already remarked[118] that sometimes there appears among the day numbers a horizontal stroke, which he transcribed by "8*" because it resembles the normal 8-sign (being a horizontal stroke with a little hook at the left end). A comparison with the calculations shows Spiegelberg's reading "8" is doubtless possible, although in some cases I had the impression that 7 or even 6 would fit better; but a clear decision between 8 and 7, which are palaeographically the only two possibilities,[119] is impossible within the framework of texts of our kind. In the translation of P I kept the reading 8, referring to it in the apparatus by "s.s." ("special sign"). The examination of all 26 cases where this 8-sign occurs shows that it always follows 10 or 20 (15-times after 10 and 11-times after 20), but I cannot see any reason why just in 18 and 28 eight should sometimes[120] be given by a different sign.

e. The Symbols of the Zodiacal Signs.—The history of the names and symbols used for the zodiacal signs in medieval and modern times is one of the most difficult chapters in the general history of astronomy. There can be little doubt that the later zodiacal symbols originated from Demotic, but it is quite beyond the scope of this edition to discuss here the questions involved.[121] I confine myself to the comparison of the typical forms given in figure 1. Except for epigraphical details, there exists agreement between P and S in the symbols for ♈, ♍, ♏, ⌐, ≈ and ♓; almost the same signs represent ♎: in P the "hill-country," in S the "rising sun." In P the sign ♉ is written as $ḥr$ "face,"[122] in S as $'nḫ$ "goat," both

___

[112] It must be remarked that 5 of the 6 exceptions to rule one and 22 of the 23 exceptions to rule two are to be found in the tablets $C_1 + C_2$; cf. p. 247.

[113] C_2 rev. I,30,31; II,1.

[114] Month 1 and the epagomenae omitted because they are covered by rule two and the remarks in section c.

[115] Day 30 is disregarded because of the remark p. 244.

[116] It is interesting to remark that the same ambiguity in the interpretation of units following tens arose in the cuneiform astronomical texts of the Seleucid age. Cf. a forthcoming notice in JAOS **61** (1941).

[117] C_1 rev. IV,3 (4 $\odot$ 1).

[118] Spiegelberg DPB p. 30.

[119] In favor of 7 there could be mentioned a hieratic form of 7 listed by Griffith [1] Pl. VI p. 22 under "abnorm. hieratic, cent. IX–VI," which is almost only a horizontal line.

[120] There occur many instances of dates 17, 18, 27, 28 written with ordinary 7 or 8 respectively.

[121] For the Demotic names of the signs see Spiegelberg [3] and a forthcoming paper by myself.

[122] Not recognized in Spiegelberg [3] p. 148. The same form occurs in the hieratic passages of the Pap. Carlsberg 1 (Lange-Neugebauer [1]).

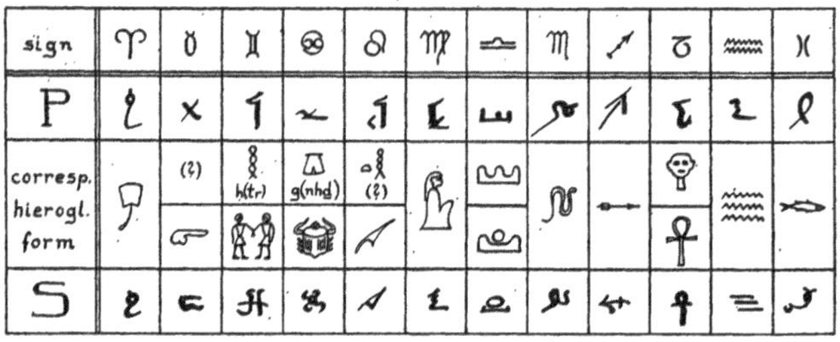

FIGURE 1

being different abbreviations of the same complete name "the goat-faced." The signs ♓ and ⊗ are given in S by the determinatives, but abbreviated to the first letter in P. As the sign for ♌ P uses the "knife" while S[123] writes the letter ḥ and some determinative (or t ?). This might be an abbreviation for ḥs3 "grim."[124] The bull is represented in S by the phallus.[125] However in P the bull is denoted by the sign ✕ (exactly as it is also used for the number 30) which probably is an abbreviated form of the phallus.[126]

One very peculiar fact is visible in P, namely representation of the zodiacal signs by simple numbers, starting with 1 meaning ♍, up to 12 meaning ♌. This notation only becomes frequent in the last columns of the papyrus, but even there the numbers are mixed with the ordinary signs without any visible rule.

The explanation of this counting of the zodiacal signs beginning with ♍ = 1 obviously lies in the fact, that in the first month of the Egyptian years covered by P the sun was mainly in the sign ♍. The accurate dates are as follows:

Egyptian date	corresp. Julian date	sun position
Augustus 14 I 1	−16 Aug. 27	♍ 2
Augustus 41 I 1	+11 Aug. 21	♌ 25

Taking into account the difference of about 4 degrees between the true vernal-point and the origin of the zodiac in our texts, we would get about ♍ 6 and ♌ 29 as the Egyptian sun positions on the New Year's day.

This kind of notation does not seem to be confined to P. The list of the zodiacal signs given in the Pap. Carlsberg 9, written after 144 A.D., starts with ♌.[127]

The starting points of the 25-year moon periods given in this text are the New Year's Days of Tiberius 14, Vespasian 1, Domitian 14, Hadrian 3 and Antoninus 7. The sun was on all those days inside ♌, except in the last instance where for the first time it entered into ⊗.[128]

Furthermore there exists a Demotic ostracon[129] where the Egyptian months are made to correspond to the zodiacal signs so that month VI[130] corresponds to ♈ and therefore month I to ♏.[131] The sun was in ♏ on the New Year's Day from about 370 to 250 B.C.[132] This brings us to the period of the calendar of Dionysius, beginning in 285 B.C., where the months are called by the zodiacal signs.[133] We have thus the following list:

Calendar of Dionysius: Eg. month I corresp. to "month ♏."

Ostraca Ð 521: month I corresp. to ♏

P: ♍ = "1"

Pap. Carlsberg 9: first sign ♌

This seems to indicate clearly that the tendency to correlate the counting of the zodiacal signs with the sun's position on the New Year's Day existed during

---

[123] Spiegelberg ([3] p. 147 and Pl. IV) erroneously states that both P and S have the "knife" for ♌. He had obviously in mind the fact that the "knife" occurs in both texts but forgot that the knife in P represents the planet "Mars" and not the zodiacal sign "lion" as in S.

[124] Cf. m3j ḥs3 "the grim looking lion" (WB III 161 and Spiegelberg [1]).

[125] Spiegelberg ([3] p. 147) says that P writes ♉ with the phallus, S with the bull, a statement for which I do not see any basis.

[126] Cf. Hess, Setne p. 192 nr. 40.

[127] Neugebauer-Volten [1] p. 385.

[128] The signs given in Neugebauer-Volten [1] p. 402 f. are erroneously one sign too short.

[129] Strassburg D 521, published in Spiegelberg [1]; cf. Müller [1] and [2] and Spiegelberg [3]. Spiegelberg [1] col. 6 gives as date "about first cent. A.D."

[130] Spiegelberg misread the months as stated in Neugebauer-Volten [1] p. 403 note 13.

[131] Details will be given in a forthcoming paper on Demotic horoscopes.

[132] This corrects the erroneous date given in Neugebauer-Volten [1] p. 403 note 14. About 250 B.C. we have practically complete correspondence between the first Egyptian month and the sign ♏ in the sense that the sun travels through this sign during this month. If we require that the sun's position at the New Year's Day lie somewhere inside ♏ then the years from 370 to 250 are possible; if we assume that only on some day of the first month the sun was inside ♏ then the years up to about 130 B.C. are admissible.

[133] The year of this calendar is not the "Egyptian year" of 365 days but the "Julian" year with an intercalation rule as in the later "Alexandrian" calendar. Cf. for this "era of Dionysius" Böckh, Sonnenkreise p. 286 ff., RE 1, 619 f.; 5, 991; 10, 1588 (the Egyptian months incorrect!); 3A, 587 f., and recently Borchardt [3] p. 8 ff.

the Demotic period. It would be of importance for chronological questions if an analogous relation could be established between the new year and the decans in older periods of Egyptian history.

f. **The Planetary Symbols.**—Both P and S agree in the arrangement of the planets (from ♄ to ☿) but are very different as far as the terminology is concerned (cf. Fig. 2). S gives the full names of the

planet	♄	♃	♂	♀	☿
P	▫	⌐ʳ	⫽	⟍	⌐ᵔ
corresp.	(?)	ₙ⌐(₃)	⫽	⫼⫼	⟍ᵔ
hierogl. form					
S					

FIGURE 2

planets [134] followed by the sign "star." P on the contrary gives only short symbols for the planets. The sign of Saturn is ▫, directly following the year numbers (cf. above p. 211) and therefore not recognized as a planetary symbol neither by Spiegelberg nor by myself. Dr. G. R. Hughes, of the Oriental Institute of the University of Chicago, however, discovered that the planetary symbols of P appear also in P. Cairo 31222 except for the additional star determinative, thus explaining the sign ▫ as the sign for Saturn. Jupiter is indicated by a short Demotic group, which I am not able to explain.[135] Mars is written as the "knife," exactly the same sign as used in S for the zodiacal sign ♌ [136]; this may represent the word dm "sharp," written in late periods as the knife,[137] although, as far as I know, its use has never been testified in connection with Mars except in P. Cairo 31222 line 8, as Dr. Hughes recognized.

The signs for Venus and Mercury in P are very curious. The number-sign "9," written exactly as everywhere else in the text, is clearly used for Venus. Perhaps the following explanation is possible: the number-word for 9 is psd but psd as a verb means "shine" or "rise"; it is especially used for stars and gods [138] and could therefore be a good cryptogram for

Venus. As a sign for Mercury the sign "string" is used in its Hieratic form; this is frequently used, e.g. in the Pap. Carlsberg 1, as a determinative for "book." The only explanation for its use here that I can propose, is the reading 'rḳ with the meaning "last," a proper designation for the innermost planet.[139]

g. **Composition of the Texts; Errors.**—It is clear in itself that our texts can not be original manuscripts but copies from other records, probably not arranged by single years but by planets. It is therefore not surprising that we are able to discover different copyist errors. Some small variations in the ductus show furthermore that neither P nor the Stobart tablets were written without interruption. This is shown in P by the suddenly increasing frequency both of the special sign for 8 in 18 and 28 [140] and of the use of numbers for the zodiacal signs.[141] Analogously small differences between the tablets A, $C_1 + C_2$ and E are visible, e.g. in the names of the planets the two final strokes never appear in A but often in $C_1 + C_2$ and E. The sign for ♐ always contains one crossline in A, in the other tablets either two or none. Tablet A consistently uses tpj for month I, $C_1 + C_2$ often ordinary 1, E ordinary 1 only once.[142] The different uses of the sign ☉ "day" have already been mentioned.[143] All this shows that the tablets A, $C_1 + C_2$ and E were written at slightly different times, although doubtless by the same scribe.

It is perhaps possible to reveal a little more about the character of the original manuscripts from which our texts were composed. S contains three errors which can only be understood as misreadings of a Demotic source (twice 1 instead of 30 and 2 instead of 7). P on the other hand substitutes 20 for 10 seven times and 10 for 20 once, errors for which no explanation can be given in Demotic characters but which are very natural in Hieratic manuscripts. That seems to indicate that the scribe of P copied a manuscript written by another scribe who used Hieratic number signs.[144] S covers 62 years and is therefore certainly not the work of the last scribe alone, if regular observations are involved as we assumed.[145]

---

[134] Discussed by Brugsch, Aeg. p. 336 ff. and Thes. I, p. 65 ff.
[135] The first sign is probably the "throw-stick" (Gardiner, Gr. T 14 in the sign-list). P. Cairo 31222 is published in Spiegelberg DD II Pl. 129.
[136] Cf. p. 246 and note 123.
[137] WB V, 448.
[138] WB I, 556 f.; cf. the pun for the "divine ennead."

[139] The original meaning of 'rḳ is "to complete"; cf. however the expression šw 'rḳ "last day" for the thirtieth day, mentioned above p. 244.
[140] Cf. p. 245. The first two cases are VII,4 and X,10; ten instances can be found in columns XVI to XXVI, 14 cases in col. XXVII to XXXVI.
[141] Cf. p. 246. The first two cases are V,15 and XI,10; column XXVII contains 4 cases and very many can be found in the last four columns.
[142] Cf. p. 244.
[143] Cf. p. 245 note 112.
[144] Even in so late a text as Pap. Carlsberg 9 (Neugebauer-Volten [1]) there occur Hieratic number signs.
[145] P covers 27 years only but is written at least 57 years after the first dates recorded (cf. p. 211).

Ordinary copying errors are of course not rare in both texts. Omissions of lines and resulting errors occur 15 times in P, 6 times in S. There remains only a very small group of errors which can not be explained in this way; their list is the following:

P	text	instead of [146]
I,1	♉	♓
III,9	(day) 11	1
IV,17	⊗	♍
V,23	(day) 20	4
VII,4	(day) 18	8
XI,2	(day) 12	18
XI,4	(day) 11	3

S	text	instead of [146]
A rev. III,23	☉ 23	☉ 30
C₂ obv. IV,29	4 29	5 7
C₂ obv. V,11	(day) 6	1

Altogether the number of errors is not higher than must be expected in copies of such dry lists of numbers. It is however easy to recognize that the scribe of S was much more careful in every respect than the copyist of P. Among 900 preserved lines P contains more than twice as many errors as S in about 1200 lines.

---

[146] These corrections are of course only approximate as far as day-numbers are concerned.

§ 6. Bibliography and Abbreviations

AN. Astronomische Nachrichten.

ÄZ. Zeitschrift für Aegyptische Sprache u. Altertumskunde.

BERGMANN, H. I. E. v. Bergmann, Hieroglyphische Inschriften gesammelt während einer im Winter 1877/78 unternommenen Reise in Aegypten. Wien, Faesy & Frick, 1879.

BÖCKH, SONNENKREISE. A. Böckh, Über die vierjährigen Sonnenkreise der Alten, vorzüglich den Eudoxischen. Berlin, Reiner, 1863.

BORCHARDT [1]. L. Borchardt, Die altägyptische Zeitmessung = vol. 1, B of Bassermann-Jordan, Die Geschichte der Zeitmessung u. d. Uhren. Berlin-Leipzig, W. de Gruyter, 1920.

BORCHARDT [2]. L. Borchardt, Ein altägyptisches astronomisches Instrument. ÄZ 37 (1899), p. 10 to 17.

BORCHARDT [3]. L. Borchardt, Versuche zu Zeitbestimmungen für die späte, griechisch-römische, Zeit der aegyptischen Geschichte (= Quellen u. Forschungen zur Zeitbestimmung d. aegyptischen Geschichte 3). Kairo, Selbstverlag, 1938.

BRUGSCH, AEG. H. Brugsch, Die Aegyptologie. Leipzig, W. Friedrich, 1891.

BRUGSCH, THES. H. Brugsch, Thesaurus inscriptionum aegyptiacarum. 1. Abtlg. Astronomische und astrologische Inschriften altaegyptischer Denkmäler. Leipzig, Hinrichs, 1883.

BRUGSCH [1]. H. Brugsch, Nouvelles recherches sur la division de l'année des anciens égyptiens, suivies d'une mémoire sur des observations planétaires consignées dans quatre tablettes égyptiennes en écriture démotique. Berlin, F. Schneider; Paris, P. Duprat. 1856.

C. Pap. Carlsberg 9 (cf. Neugebauer-Volten [1]).

CHASSINAT [1]. E. Chassinat, Le temple d'Edfou III = MMAF 20, 1928.

CHATLEY [1]. H. Chatley, Notes on Ancient Egyptian Astronomy. The Observatory 62, p. 100–104 (1939).

CHATLEY [2]. H. Chatley, Ancient Egyptian Astronomy. Nature 143, p. 336 (1939).

CLEMENS ALEXANDR. Strom. ed. O. Stählin (im Auftrage der Kirchenväter-Commission d. Kgl. Preuss. Akad. d. Wiss.) vol. II. Stromata I–VI. Leipzig, Hinrichs, 1906. transl. Fr. Overbeck, Die Teppiche, Basel, Schwabe & Co., 1936.

CUMONT [1]. Fr. Cumont, L'Egypt des Astrologues. Bruxelles, 1937.

DIELS VS. H. Diels, Die Fragmente der Vorsokratiker. 5. ed. (by W. Kranz). Berlin, Weidmann, 1934 to 1937.

DPB. Demotische Papyrus aus den Königlichen Museen zu Berlin. Herausgegeben von der Generalverwaltung. Giesecke & Devrient, Berlin-Leipzig, 1902.

DREYER [1]. J. L. E. Dreyer, History of the Planetary System from Thales to Kepler. Cambridge Univ. Press, 1906.

GARDINER, GR. A. H. Gardiner, Egyptian Grammar. Oxford, Clarendon Press, 1927.

GHG. Grenfell-Hunt-Goodspeed (cf. Tebt. Pap.).

GINZEL, CHRON. F. K. Ginzel, Handbuch der mathematischen und technischen Chronologie. Leipzig, Hinrichs, 1906–1914.

GRIFFITH [1]. F. Ll. Griffith, Meroitic Studies. JEA 3 (1916) p. 22 ff.

GRIFFITH [2]. F. Ll. Griffith, The Old Coptic Horoscope of the Stobart Collection. ÄZ 38 (1900) p. 71–85.

GUNDEL, HERMES TRISMEGISTOS. W. Gundel, Neue astrologische Texte des Hermes Trismegistos, Abh. d. Bayerischen Akad. d. Wiss. Phil.-hist. Abt., NF 12 (1936).

HARPER, LETTERS. See Waterman, R. C.

HEATH, ARISTARCH. Sir Thomas Heath, Aristarchus of Samos. Oxford, 1913.

HESS, SETNE. J.-J. Hess, Der demotische Roman von Stne ḫa-m-us. Leipzig, Hinrichs, 1888.

JAOS. Journal of the American Oriental Society.

JEA. The Journal of Egyptian Archaeology.

KENYON, Cat. Pap. Br. Mus. I. F. G. Kenyon, Greek Papyri in the British Museum. Catalogue with texts. I. London, 1893.

KUGLER, SSB. F. X. Kugler, Sternkunde u. Sterndienst in Babel. Münster, Aschendorff, 1907 ff.

LANGE-NEUGEBAUER [1]. H. O. Lange and O. Neugebauer, Papyrus Carlsberg No. 1, ein hieratisch-demotischer kosmologischer Text. Det Kgl. Danske Vid. Selskab. Hist.-fil. Skrifter vol. 1, no. 2. Copenhagen, 1940.

LYONS [1]. H. Lyons, Two notes on land-measurement in Egypt. JEA 12 (1926) p. 242 f.

MACROBIUS, SOMN. SCIP. Macrobius, Commentarium in somnium Scipionis, ed. Eyssenhardt. Leipzig, Teubner, 1893, p. 476 ff.

MGMN. Mitteilungen zur Geschichte der Medizin und der Naturwissenschaften.

MMAF. Mémoires publiés par les membres de la Mission Archéologique Française au Caire.

MÖLLER, Pal. III. Georg Möller, Hieratische Palaeographie Bd. III. Von der 22.-ten Dynastie bis zum 3.-ten Jahrh. n. Chr. 2-te Aufl. Leipzig, Hinrichs, 1936.

MÜLLER [1]. W. Max Müller, Zu dem neuen Strassburger astronomischen Schultext. OLZ 5 (1902), Sp. 135 f.

MÜLLER [2]. W. Max Müller, Zur Geschichte der Tierkreisbilder in Aegypten. OLZ 6 (1903), Sp. 8 f.

MVAG. Mitteilungen der Vorderasiatischen Gesellschaft.

NEUGEBAUER [1]. O. Neugebauer, Egyptian Astronomy. Nature 143, p. 115 f. (1939).

NEUGEBAUER [2]. O. Neugebauer, Ancient Egyptian Astronomy. Nature 143, p. 765 (1939).

NEUGEBAUER [3]. O. Neugebauer, Über eine Methode zur Distanzbestimmung Alexandria-Rom bei Heron. Kgl. Danske Vidensk. Selsk., Hist.-fil. Medd. 26, 2 (1938).

NEUGEBAUER-VOLTEN [1]. O. Neugebauer-A. Volten, Untersuchungen zur antiken Astronomie IV. Ein demotischer astronomischer Papyrus (Pap. Carlsberg 9). QS B 4 (1938), p. 383–406.

P. V. NEUGEBAUER, T. A. CHR. P. V. Neugebauer, Tafeln zur astronomischen Chronologie. III. Hilfstafeln zur Berechnung von Himmelserscheinungen. Leipzig, Hinrichs, 1922.

P. V. NEUGEBAUER [1]. P. V. Neugebauer, Genäherte Tafeln für Sonne und Planeten. Astronomische Nachrichten 248 Nr. 5937, p. 161 ff. (1933).

P. V. NEUGEBAUER [2]. P. V. Neugebauer, Hilfstafeln zur technischen Chronologie. Teil II, Das Wandeljahr und das gebundene Mondjahr. Astronomische Nachrichten 261 (Nr. 6261), p. 377 ff. (1937).

OEFELE [1]. [F.v.] O[efele], Review of DPB in MGMN 1 (1902), p. 327–330.

OEFELE [2]. F. v. Oefele, Die Angaben der Berliner Planetentafeln P 8279 verglichen mit der Geburtsgeschichte Christi im Berichte des Matthäus. MVAG 8, Heft 2 (1903).

OLZ. Orientalische Litteratur-Zeitung.

P. Papyrus Berlin P 8279 (published in Spiegelberg DPB).

PAPPUS in Almag. VI ed. Rome. A. Rome, Commentaires de Pappus et de Théon d'Alexandria sur l'Almageste. Tome I. Pappus d'Alexandria Commentaire sur les livres 5 et 6 de l'Almageste. Studi e testi 54. Roma, Biblioteca Apostolica Vaticana, 1931.

POGO [1]. A. Pogo, Egyptian water clocks, Isis 25 (1936), p. 403 to 425.

QS. Quellen und Studien zur Geschichte der Mathematik, Astronomie und Physik.

r.b.c. restoration by calculation (explanation see p. 211).

r.b.i. restoration by interpolation (explanation see p. 211).

RE. Real-Encyklopädie der classischen Altertumswissenschaften, herausgeg. v. Pauly-Wissowa.

rev. reverse.

SCHNABEL [1]. P. Schnabel, Kidenas, Hipparch und die Entdeckung der Präzession. ZA 37 (1927), p. 1 to 60.

SCHOTT [1]. A. Schott, (Critical review of) W. Gundel, Neue astrologische Texte des Hermes Trismegistos. QS B 4 (1938), p. 167–178.

SCHRAM, TAFELN. R. Schram, Kalendariographische und chronologische Tafeln. Leipzig, Hinrichs, 1908.

SCOTT, HERMETICA. W. Scott, Hermetica, the ancient Greek and Latin writings which contain religious or philosophic teaching ascribed to Hermes Trismegistus. 4 vols. Oxford, Clarendon Press, 1924 to 1936.

SETHE, ZEITRECHNUNG. K. Sethe, Die Zeitrechnung der alten Aegypter im Verhältnis zu der der andern Völker. Nachrichten d. K. Ges. d. Wissensch. zu Göttingen, Phil.-hist. Kl. 1919, p. 287–320; 1920, p. 28–55, 97–141.

SLOLEY [1]. R. W. Sloley, Primitive methods of measuring time with special reference to Egypt. JEA 17 (1931), p. 166–178.

SPIEGELBERG DD II. W. Spiegelberg, Die demotischen Denkmäler II. Die demotischen Papyrus. Catalogue général des antiquités égyptiennes du Musée du Caire 40, Strassburg 1906.

SPIEGELBERG DPB. W. Spiegelberg, Demotische Papyrus aus den Königlichen Museen zu Berlin, Herausgegeben von der Generalverwaltung. Leipzig u. Berlin, Giesecke & Devrient, 1902.

SPIEGELBERG, GR. W. Spiegelberg, Demotische Grammatik. Heidelberg, Winter. 1902.

SPIEGELBERG [1]. W. Spiegelberg, Ein aegyptisches Verzeichnis der Planeten und Tierkreisbilder. OLZ 5 (1902), Sp. 6–9.

SPIEGELBERG [2]. W. Spiegelberg, Ein neuer astronomischer Text auf einem demotischen Ostrakon. OLZ 5 (1902), Sp. 223–225.

SPIEGELBERG [3]. W. Spiegelberg, Die ägyptischen Namen und Zeichen der Tierkreisbilder in demotischer Schrift. ÄZ 48 (1910), p. 146–151.

SPIEGELBERG [4]. W. Spiegelberg, Varia. ÄZ 53 (1917), p. 91–115.

s.s. special sign (cf. explanation p. 245).

STOBART [1]. Egyptian Antiquities collected on a voyage made in Upper Egypt in the years 1854 & 1855 and published by Rev^d. H. Stobart M.A. Queen's College Oxford. Paris chez Benj. Duprat, Rue du Cloitre-St. Benoit 7. Berlin chez F. Schneider & Co. Unter den Linden 19. 1855. Värsch & Happe. lithogr. fac.-sim. under the direction of Dr. H. Brugsch, Berlin.

T. Tebt. Pap. II, 274.

TEBT. PAP. II. The Tebtunis Papyri. Part II ed. by B. P. Grenfell, A. S. Hunt, E. J. Goodspeed. London, Frowde, 1907 (= Univ. of California Publications, Graeco-Roman Archaeology vol. II).

THEO SMYRN. Theonis Smyrnaei philos. Plat. expos. rerum math. ad legendum Platonem utilium, ed. E. Hiller. Leipzig, Teubner, 1878.

Theonis Smyrnaei Platonei liber de Astronomia, ed. Th. Martin. Paris, 1849.

Théon de Smyrne, trad. J. Dupuis. Paris, Hachette, 1892.

THOMPSON, REP. R. C. Thompson, The reports of the magicians and astrologers of Niniveh and Babylon in the British Museum. London, Luzac, 1900 (= Luzac's Semitic text and translation series vol. 6 and 7).

VOGT [1]. H. Vogt, Versuch einer Wiederherstellung von Hipparchs Fixsternverzeichnis. AN 224 (1925), 17–54.

WALLACE [1]. Edith O. Wallace, The notes on philosophy in the commentary of Servius on the Eclogues, the Georgics and the Aeneid of Vergil, Columbia Univ. Press, 1938.

WATERMAN, RC. L. Waterman, Royal correspondence of the Assyrian Empire. University of Michigan Studies, Humanistic Series, vol. 17–20. Ann Arbor, Univ. of Michigan Press, 1930–1936.

WB. Erman-Grapow, Wörterbuch der Aegyptischen Sprache. Leipzig, Hinrichs, 1926–1931.

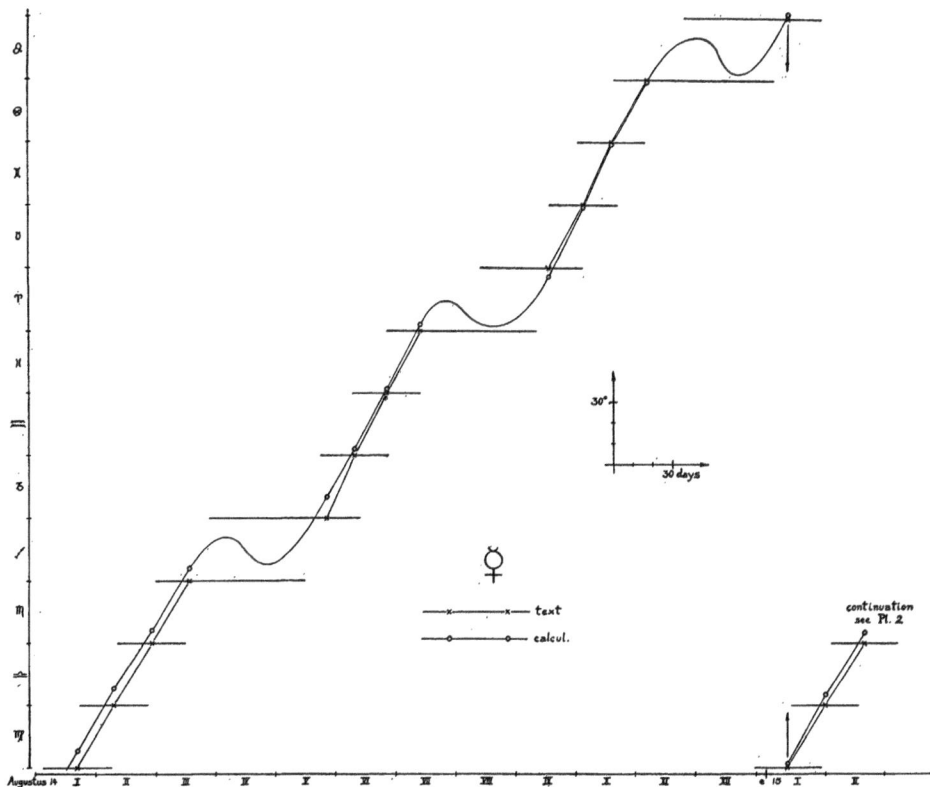

PLATE 1

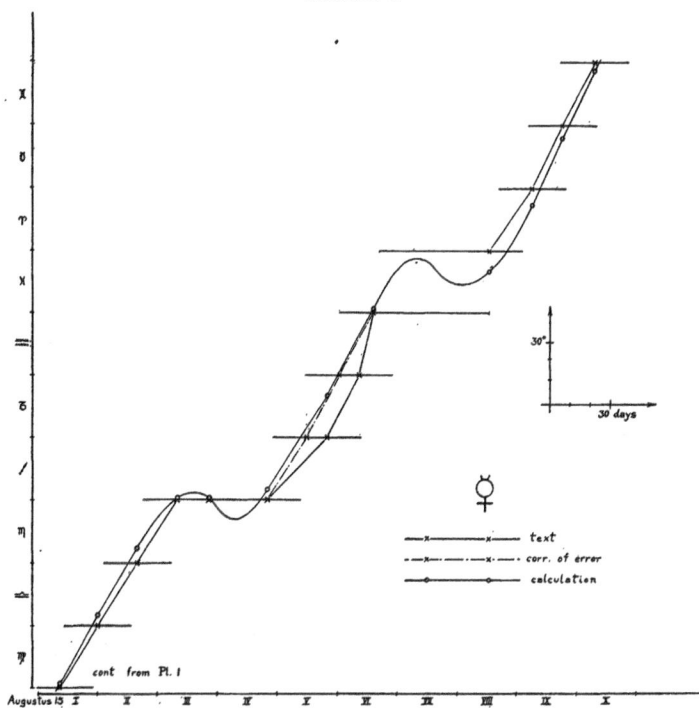

PLATE 2

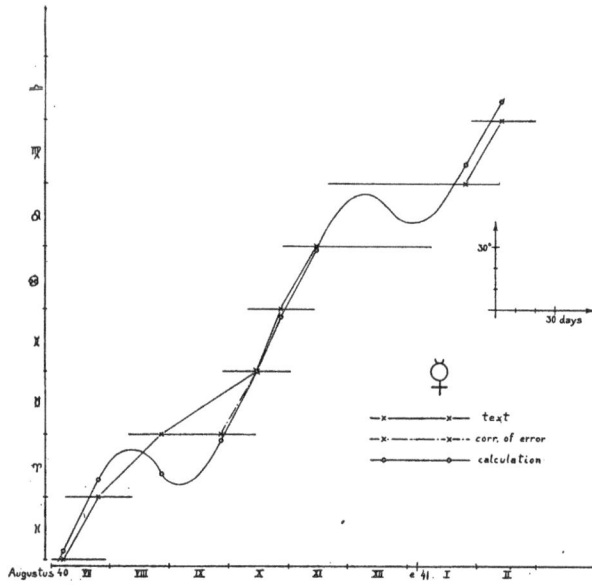

PLATE 3

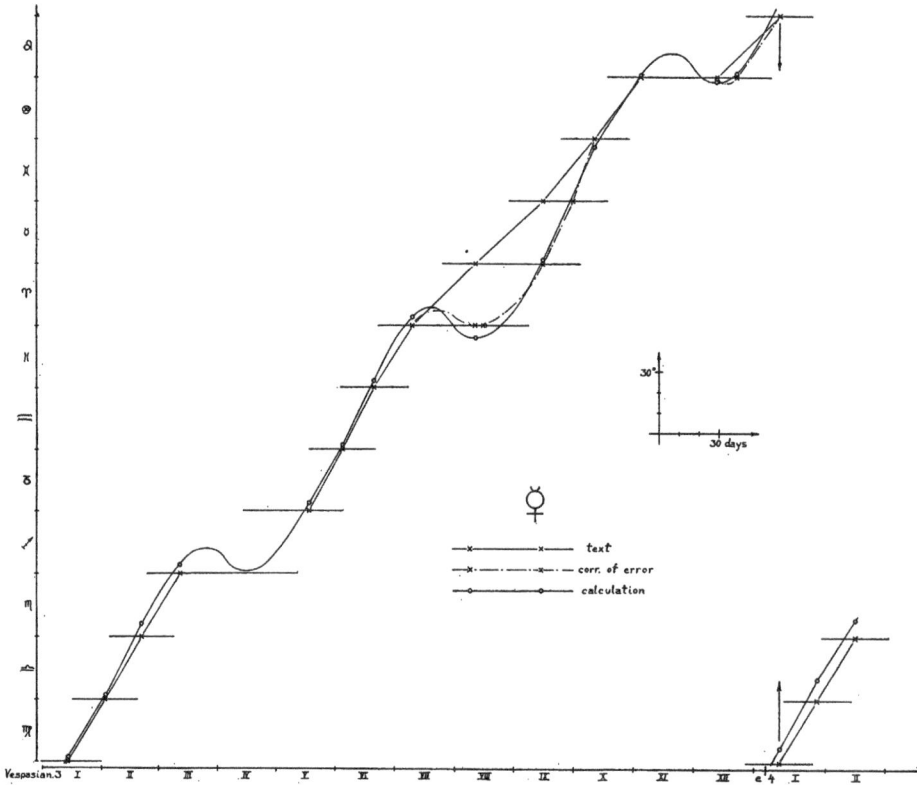

PLATE 4

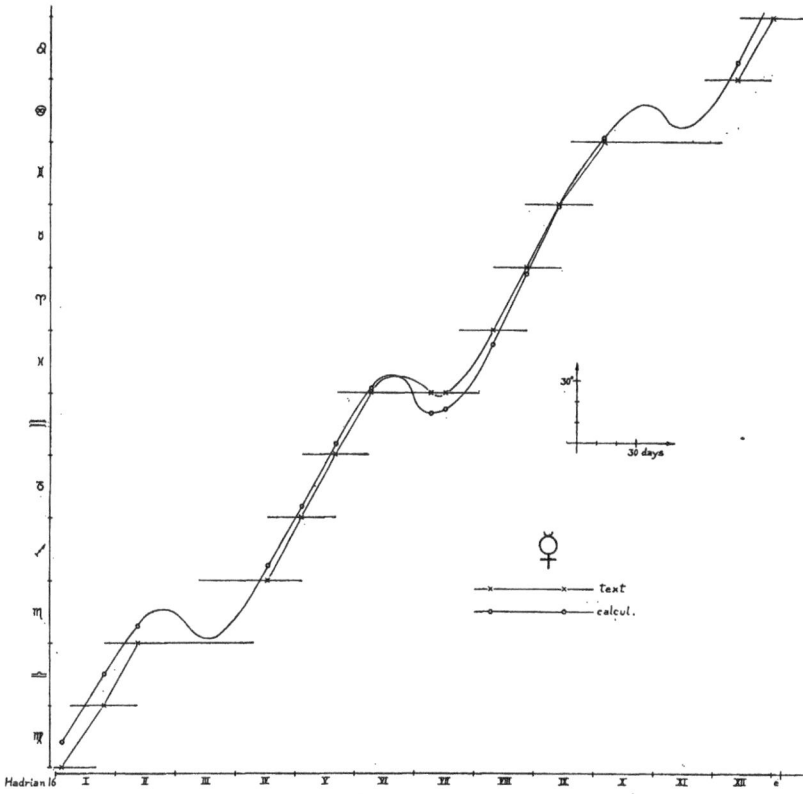

PLATE 5

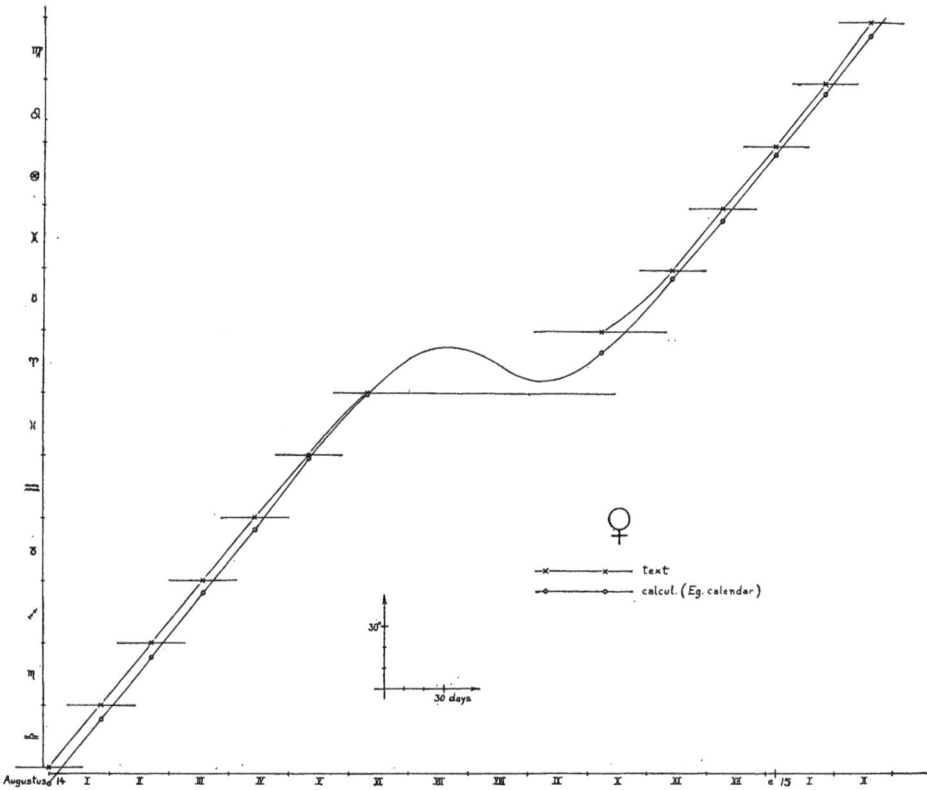

PLATE 6

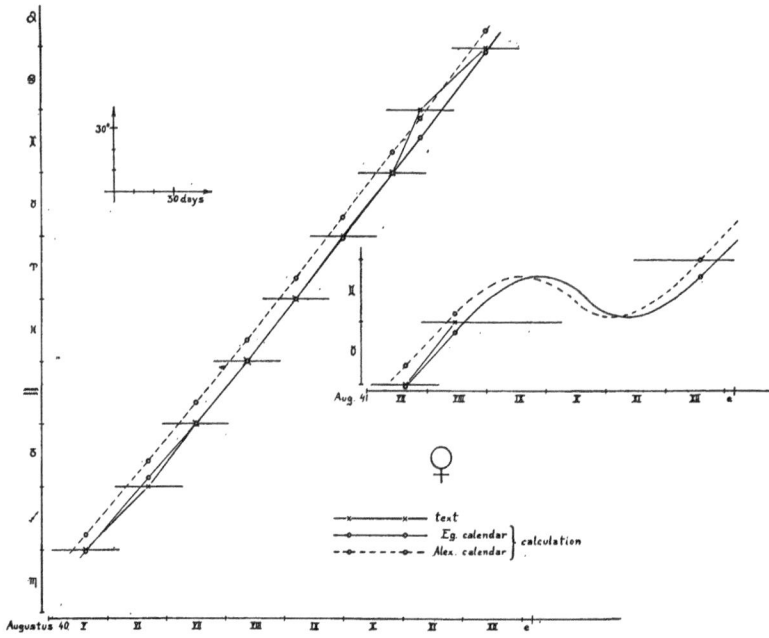

PLATE 7

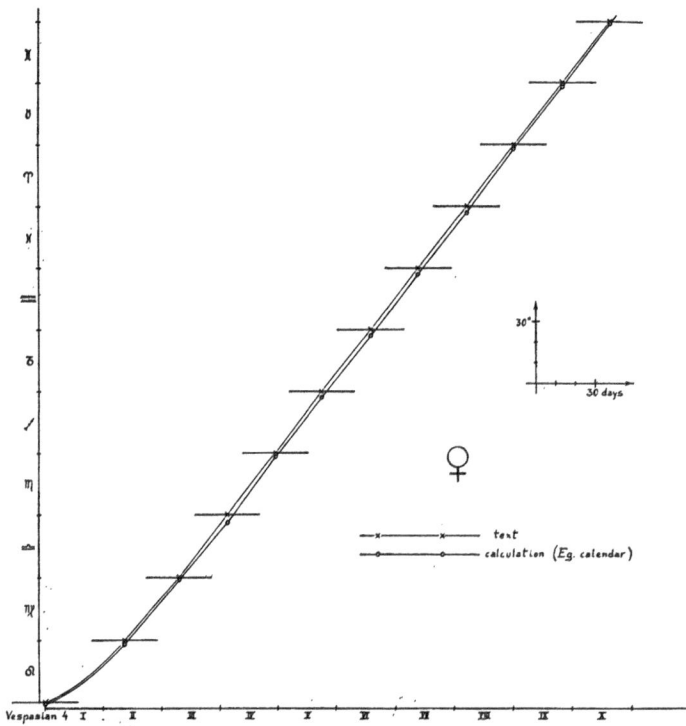

PLATE 8

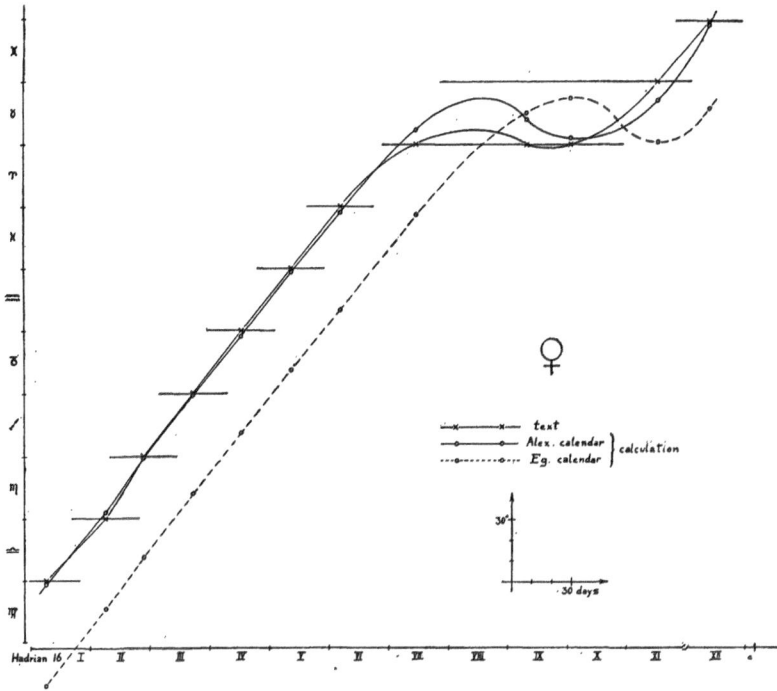

PLATE 9

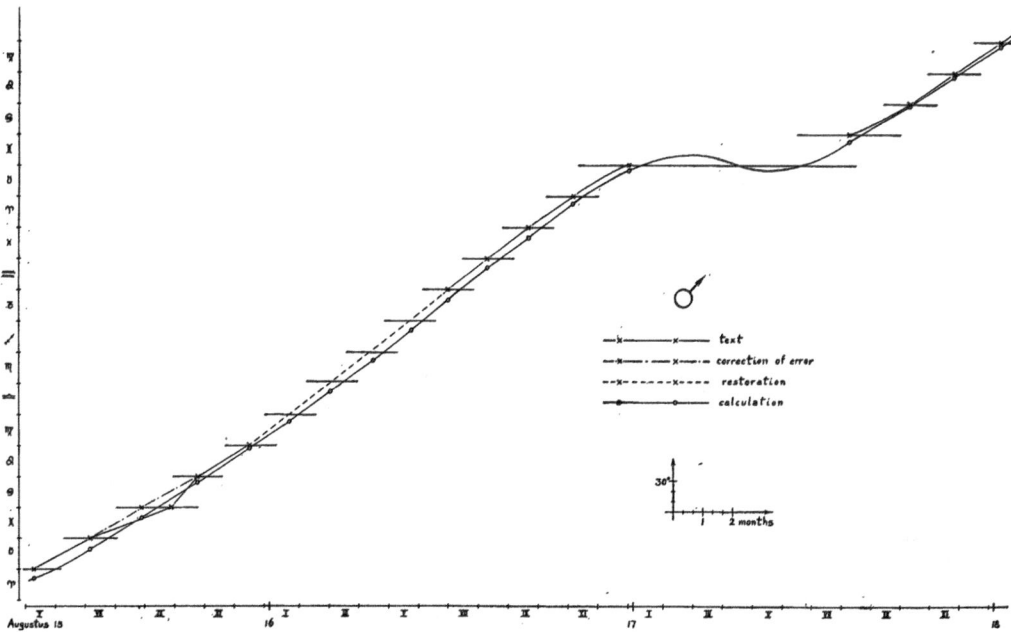

PLATE 10

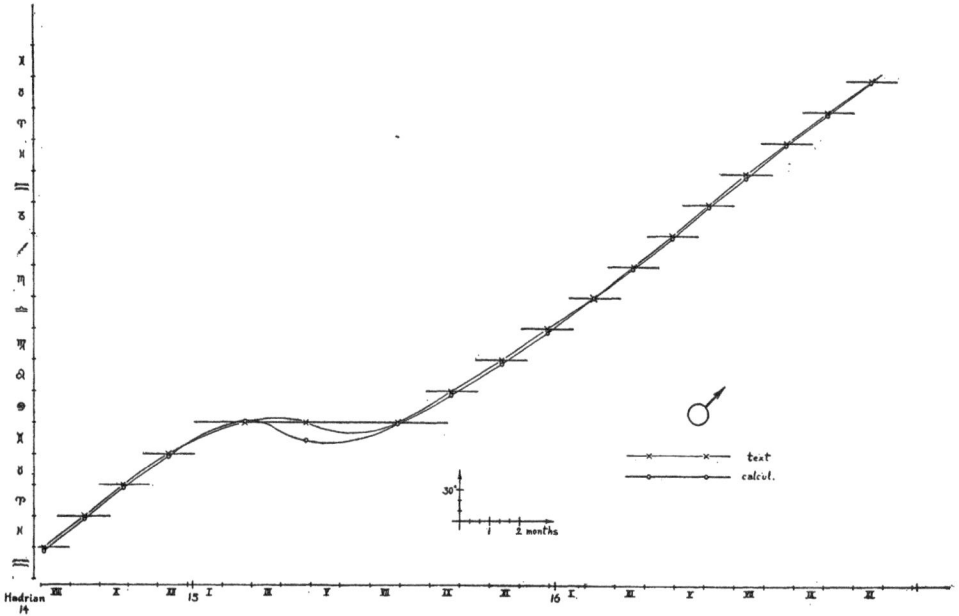

PLATE 11

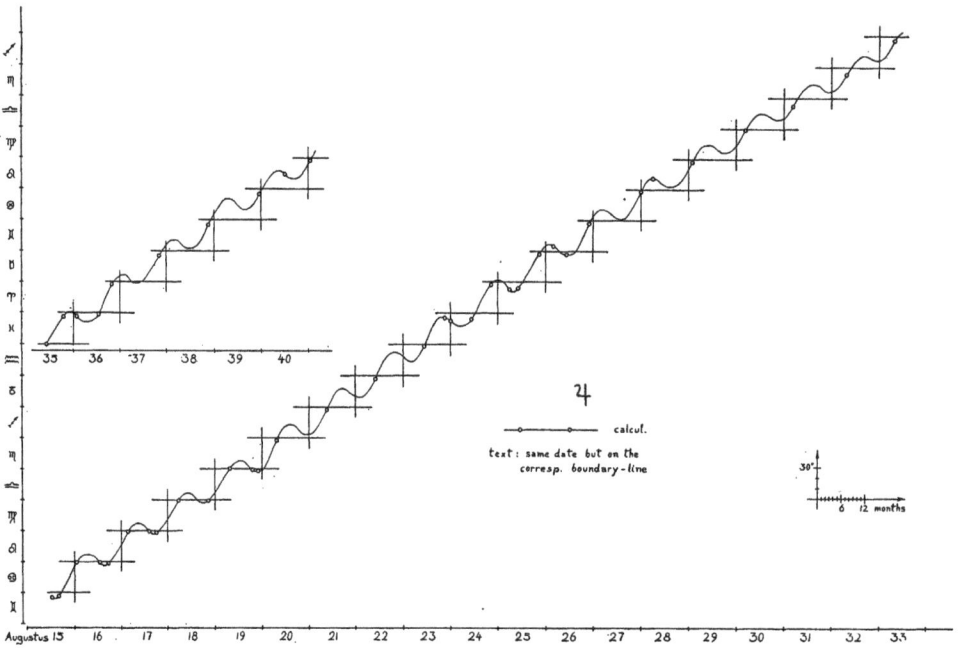

PLATE 12

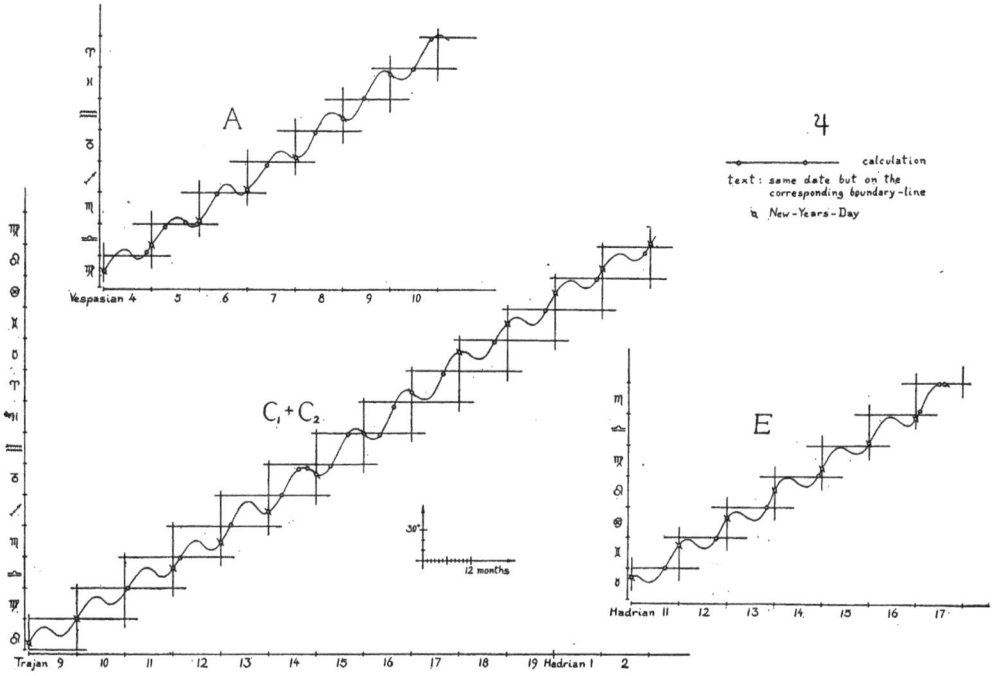

PLATE 13

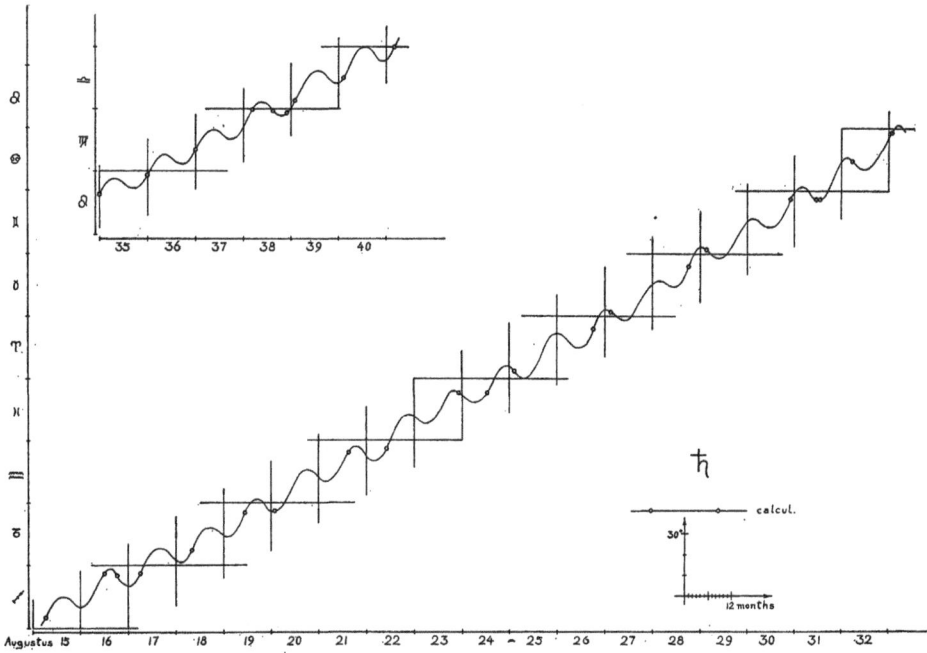

PLATE 14

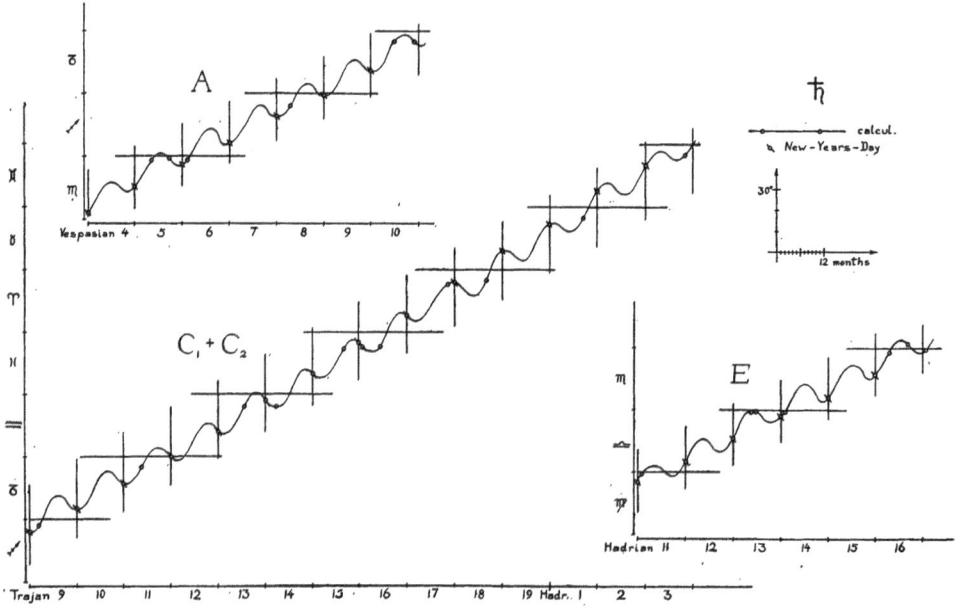

PLATE 15

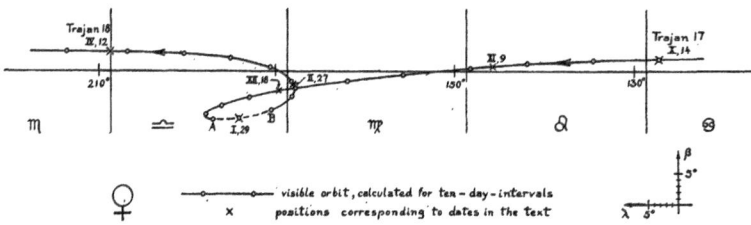

PLATE 16

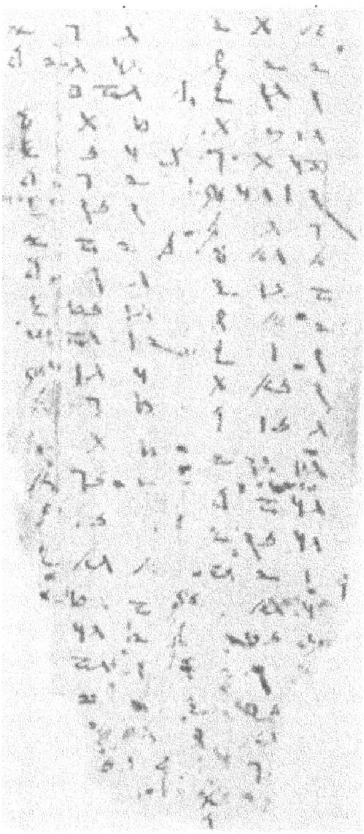

PLATE 17

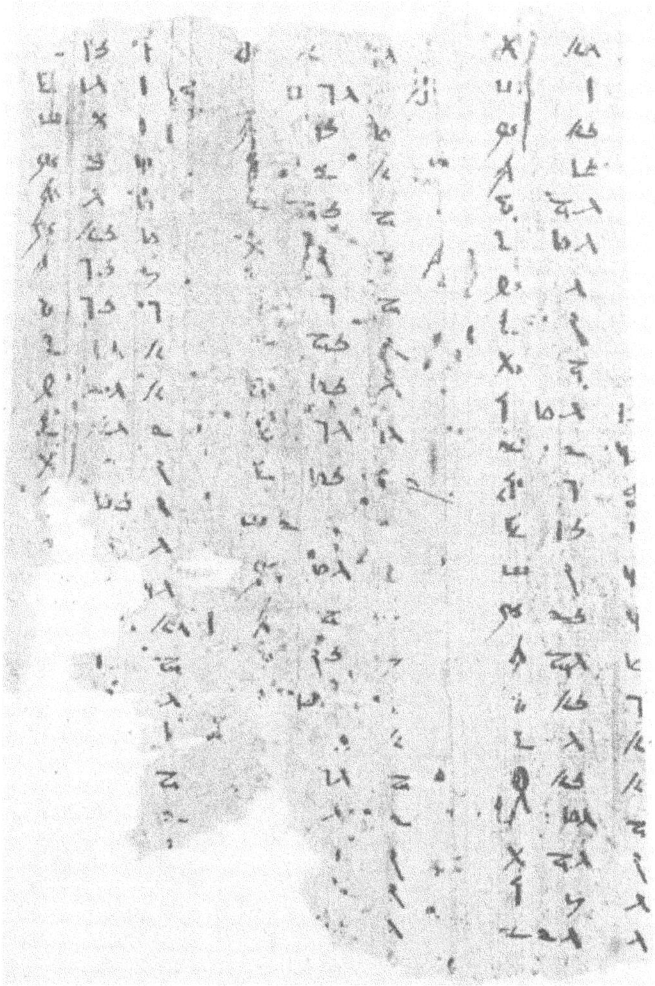

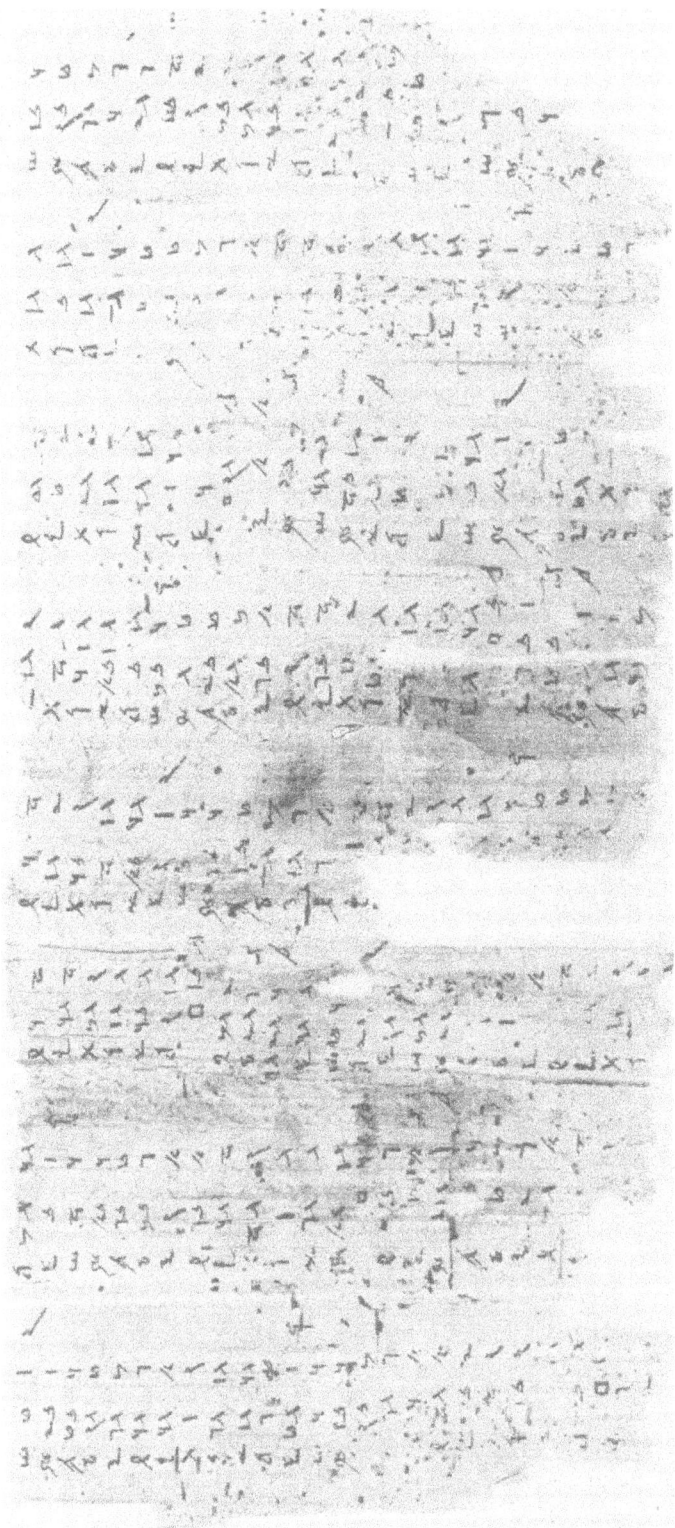

PLATE 18

XX   XIX   XVIII   XVII   XVI   XV   XIV

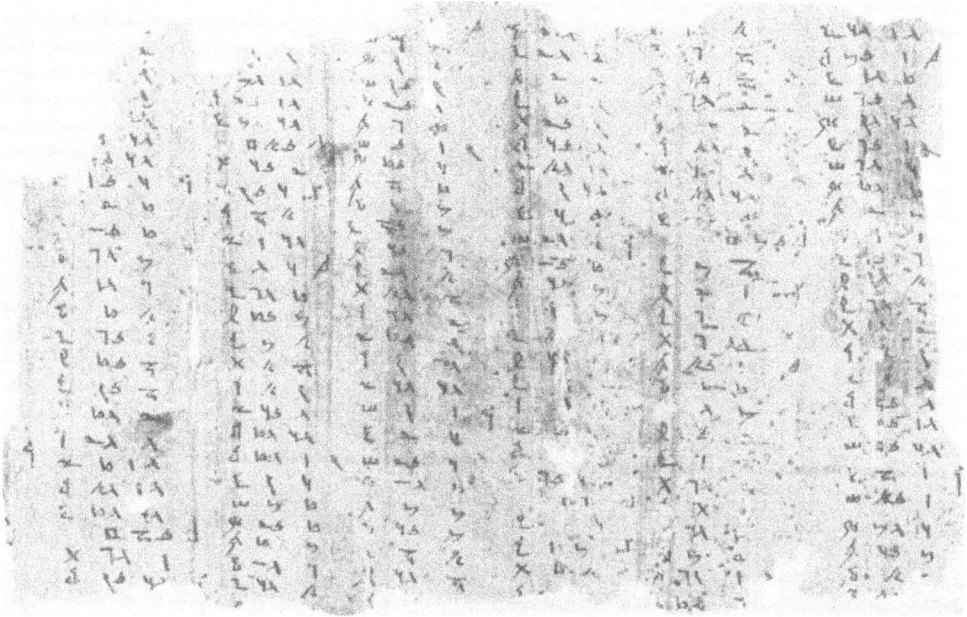

PLATE 19

XXIV   XXIII   XXII   XXI   XX

PLATE 20

XXX　　　XXIX　　　XXVIII　　　XXVII　　　XXVI　　　XXV

PLATE 21

PLATE 22

rev.   obv.

A

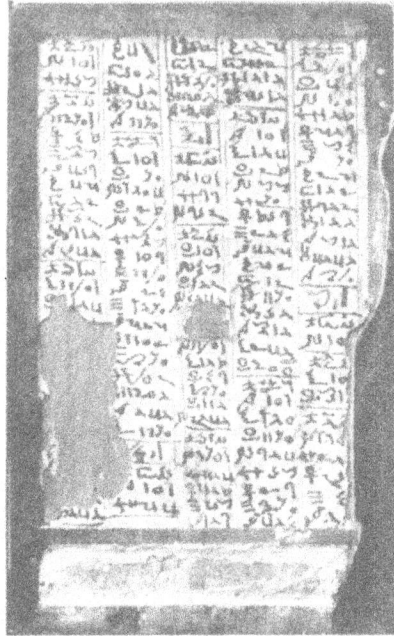

PLATE 23

rev.   obv.

C₁

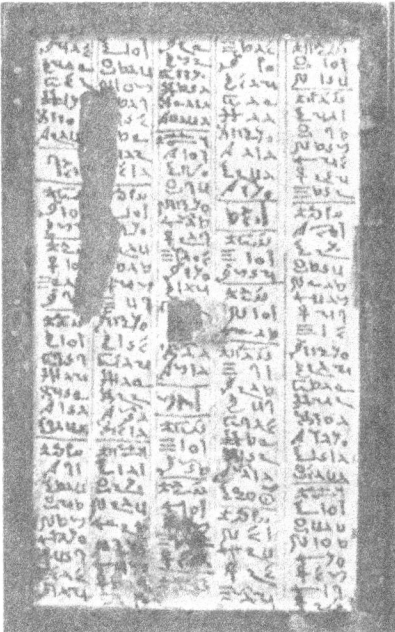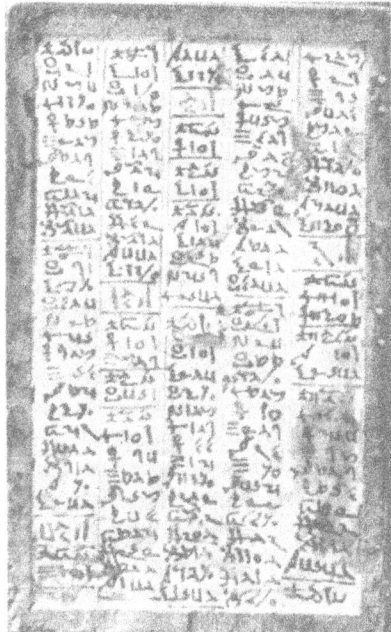

PLATE 24

rev. obv.

C₂

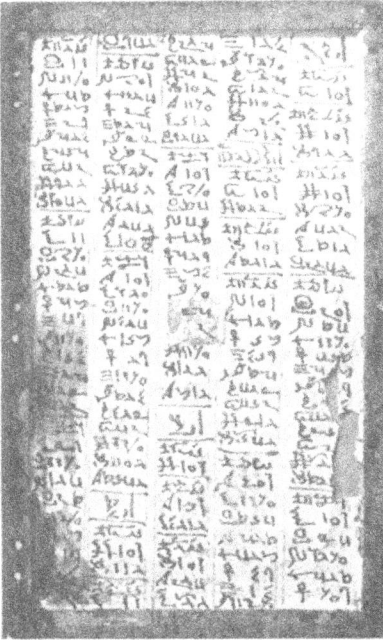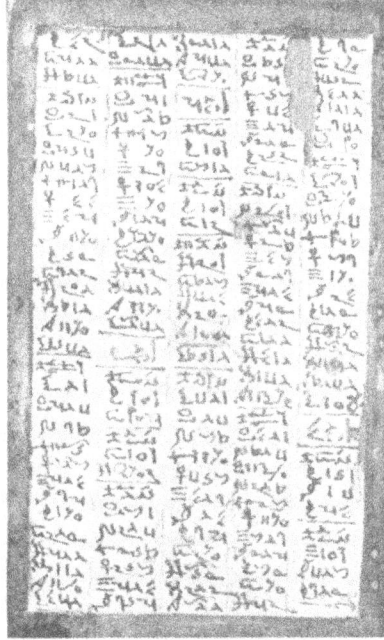

PLATE 25

rev. obv.

E

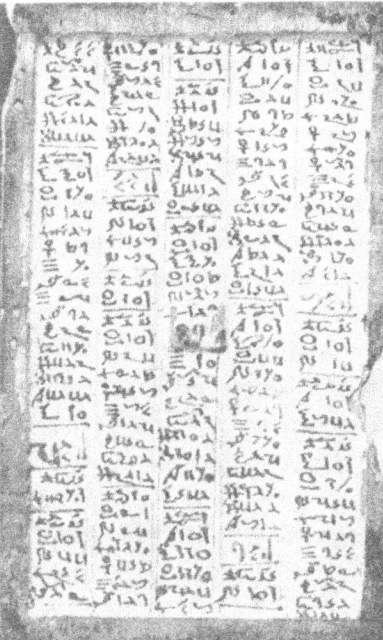

PLATE 26

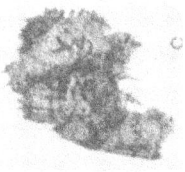

VI″

c

5

PLATE 27

I′   II′   III′   IV′

5   10   15   20   25

d

VI

5   10

a

VII

25

b

ON SOME ASTRONOMICAL PAPYRI AND RELATED PROBLEMS
OF ANCIENT GEOGRAPHY

By O. Neugebauer

F. E. Robbins published in 1927 "A new astrological treatise" which is now contained in the third volume of the "Michigan Papyri"[1] as No. 149. This interesting text gives us (in addition to other material) information about the ἀναφοραί, the time of oblique ascension of the zodiacal signs, a concept the importance of which for the ancient theory of the "climata" was first fully understood by E. Honigmann.[2] The same author felt the close relationship between the Greek methods of dealing with this problem and certain features in the Babylonian astronomy of Seleucid times,[3] but, misled by some incomplete or even erroneous information, he did not quite succeed in establishing this relationship to its full extent. I intend in this paper to complete these investigations as far as Greek astronomy is concerned.[4]

The same problem of the ἀναφοραί is dealt with in the Pap. Michigan III, 151, a fact not recognized by the editors because of the bad mutilation of this fragment. I shall give a restoration and commentary of this text as far as its state of preservation permits.

A third astronomical fragment, Pap. Michigan III, 150, refers to the moon. A few more details which still can be gathered from this text might be of interest for the restoration of manuscripts of the same kind.

The notation applied in the following for writing numbers and fractions is consistently sexagesimal: the sign " ; " separates integers from fractions and the sign " , " different powers of sixty. Thus 1,30 means 90, but 1;30 represents $1\frac{30}{60} = 1\frac{1}{2}$.

Finally, I must express my warmest gratitude to Dr. F. E. Robbins and Dr. H. C. Youtie for frequent careful investigation of the originals in the papyrus collection of the University of Michigan.

§ 1. Pap. Mich. III, 150

1. The content of this small "badly preserved" fragment is given by the following transliteration:[5]

1.	] D	6;[.]ʰ	♓	7;21	[I]
	] N	2;40	♈	6;35	II
	] N	12;0	♉	6;6	[II or III]
	]9 D	11;6	♊	5;56	[III or IV]
5.	] D	0;20	♋	5;3	IV
	]7 N	[.];10	♌	5;40	[V]
	·] D	[.;.]	♍	5;2	VI
	] N	[.;.]	♎	4;4	VII
	] D	11;0	♏	2;49	VIII
10.	] D	3;20	♐	1;19	IX
	] D	2;20	♐	29;21	X
	] D	5;20	♑	27;55	XI
	] N	6;[.]	♒	26;43	[XII]

D and N here mean "day" and "night" respectively, ʰ stands for ὥρα and the Roman numerals represent the Egyptian months in their regular order, "I" corresponding to "Thoth," "II" to "Phaophi" etc. No left hand or right hand margin is preserved.

2. First, we shall prove the correctness of the conjecture of the editor that "the purpose of the table was to give the positions of the sun on the days of successive new (or full) moons," although with the slight modification that the positions do not refer to the sun but to the moon in opposition to the sun, or in other words, to the full moons.

This proof is very simple. Twelve steps lead from ♓ 7;21 to ♒ 26;43, thus covering 5,49;22 degrees. One step corresponds therefore to 5,49;22 : 12 = 29;6,50°. On the other hand, on the average the sun travels daily ca. 0;59,9° and one synodical month is about 29;32 days long.[6] The average path of the sun during one synodic month is therefore 0;59,9 · 29;32 ≈ 29;6,54° or almost exactly the amount derived from our text.[7]

In order to determine whether the text refers to new moons or full moons, we only have to consider the change in the distances from month to month as given by the text:

---

[1] See Robbins [1] in the Bibliography at the end.
[2] Honigmann [1].
[3] M. P. III p. 312 ff.
[4] For the general importance of this problem see Neugebauer [1].
[5] I owe the following additions to the edition in M. P. III to Dr. H. C. Youtie: line 2: θ[ωθ] is an error in reading, the papyrus has φ[αωφι]; line 4:]θ̄; line 6: ε̄ or ϛ̄ or ζ̄ (as shown in the text ϛ̄ is the correct reading); line 9: read φϕ[ρμουθι] in place of [φαρμουθι]; line 13: read ρ̣υκτ(ος) instead of [...] and Με[σορη in place of [Μεσορη].

[6] The Pap. Michigan III, 149 gives for the sun's velocity 0;59,8,16° per day but Babylonian astronomy from Seleucid time uses 0;59,9 (cf. e.g. Schnabel [1] p. 39, but this paper contains many errors in the commentary as well as in details and in the main thesis). For the length of the synodic month see e.g. Geminus, cap. 18 (ed. Manitius p. 200 ff.).
[7] It is a lapsus calami of the editor when he says (p. 118) that the sun travels "an average of somewhat less (!) than 29° during the moon's synodic period."

	text	difference			text	difference			text	= day	difference
1.	♓ 7;21				♎ 4;4	29;2			N [8;.]	[20;.]	[14;.]
	♈ 6;35	29;14			♏ 2;49	28;45			D 11;0	11;0	[15;.]
	♉ 6;6	29;31	10.		♐ 1;19	28;30	10.		D 3;20	3;20	16;[.]
	♊ 5;56	29;50			♐ 29;21	28;2			N 2;20	14;20	11;0
5.	♋ 5;3	29;7			♑ 27;55	28;34			D 5;20	5;20	15;0
	♌ 5;40	30;37			♒ 26;43	28;48			N 6;[.]	18;[.]	13;[.]
	♍ 5;2	29;22									

This shows immediately that the movement during the first six months is faster than during the second part of the year, having a secondary maximum in ♊, its minimum in ♐. However, at the beginning of our era the sun's velocity has its minimum in ♊,[8] its maximum in ♐, or in other words just at the points opposite to the places given in the text. This is only possible if the text does not refer to the conjunctions of the moon and the sun but to the oppositions, i.e. to the full moons.[9]

This conclusion is furthermore confirmed by the following consideration. The second column of the text gives Egyptian months from I to XII.[10] Therefore it is highly probable that the first position given in the first column corresponds to the beginning of the year. But in Roman times the sun never stands in the sign of ♓ at the beginning of the Egyptian-Hellenistic calendar, but just opposite in ♌ or ♍. Again only full moons agree with the positions given in the text.

3. After having established the general character of the text we now discuss the dates given in the first part of the first column. The hours of "day" or "night" are abbreviated in the papyrus by ωρ ͺ. Unfortunately it seems that this symbol is unknown from elsewhere. It might mean ὥρα or perhaps more precisely ὥρα ἰσημερίνη because ὥρα alone would be rather inconvenient in an astronomical table, leaving it open whether these "hours" should be interpreted as seasonal hours or as equinoctial hours. For the sake of simplicity we shall assume the latter case, which is at any rate sufficient to restore roughly the gaps in this part of the papyrus, as shown in the following table:

	text	= day	difference
1.	D 6;[.]^h	6;[.]^h	
	N 2;40	14;40	8;[.]^h
	N 12;0	24;0	9;20
	D 11;6	11;6	11;6
5.	D 0;20	0;20	13;14
	N [3;]10	[15;]10	[14;]50
	D [6;.]	[6;.]	[15;.]

[8] Cf. e.g. Ptolemy, Almagest III, 4 (ed. Heiberg I, p. 237, 9–11) Apogee: ♊ 5;30°.

[9] The maximum in ♌ must reflect the position of the apogee of the moon. The reading εμ in line 6 cannot be replaced by anything else, as Prof. Youtie kindly informed me.

[10] The repetition of II, III or IV is due to the fact that there must always exist years containing 13 full or new moons (cf. e.g. Neugebauer-Volten [1] for the Egyptian moon calendar).

The comparison of the first and the last line of this table shows a time difference of [354 days] + 12 hours, which agrees with the average length of twelve months (of 29.53 days each) i.e. 354 days + 9 hours.

It is interesting that the reexamination of the papyrus by Prof. Youtie gave in lines 4 and 6 remainders of the dates which must have preceded the hours. In line 4]θ̄ is visible i.e. 9 or 19 or 29. The traces in line 6 could be read as θ̄, ε̄ or ε̣. The last possibility is excluded by the fact that the date of the full moon cannot drop from the 9th to 5th during two months only. Assuming morning epoch in the counting of dates we should get as time difference from month m day 9 daytime 11;6 in line 4 to month $m + 2$ day 9 night 3;10 in line 6 a difference of 60 days and 4;4^h, which is too much for two months. If we however assume evening epoch, then this difference becomes 59^d + 4;4^h only. On the other hand the reading 7 in line 6 in the case of morning epoch gives a time interval of 58^d + 4;4^h; in the case of evening epoch however 57^d + 4;4^h only, which is too short a time. In other words, the reading 9 requires the assumption of evening epoch, the reading 7 in line 6 requires morning epoch. The existence of the evening epoch in Egyptian dates was strongly contested by Sethe against Ed. Meyer.[11] The most plausible restoration is therefore

line 4.   [month m day .]9   day 11;6
     5.   [month $m + 1$ day .8]   day 0;20
     6.   [month $m + 2$ day .]7   night [3;]10

4. One could attempt to date this document exactly, using the positions and hours given in the text. There are, however, so many possibilities which fit these positions equally well that a choice could be made only by ascribing to some of the numbers a reliability which cannot be granted because of the general character of the text, even if we disregard the open problem as to which point of the ecliptic should be assumed as origin. Furthermore, it is obvious that the positions recorded are not the result of observations (full moons during day time!) but of some theoretical calculations the exactitude of which we are not able to judge. We must therefore confine ourselves to the restorations given above.

[11] Sethe [1] p. 130 f.

§ 2. General Remarks about the "Anaphorai" and their Measurement

5. We suppose the ecliptic subdivided into twelve parts of 30 degrees each, beginning with the vernal point and proceeding in the direction of the annual movement of the sun. These twelve parts may be called the "zodiacal signs" even when the vernal point does not coincide with the beginning of the sign "aries" but is located for instance on the eighth degree of this sign. We shall emphasize the distinction between the two overlapping subdivisions only when necessary to avoid confusion.

The problem of the "Anaphorai" or oblique ascensions consists in the determination of the time intervals α_1, α_2, $\cdots$, α_{12} required by the first, second, $\cdots$, twelfth sign to rise above a given horizon. These time intervals are obviously not equally long because of the inclination of the ecliptic to the plane of the daily rotation, the celestial equator. The problem of determining these α's occurs therefore wherever arcs of the ecliptic are to be compared with arcs on the equator—a problem which plays for instance an important rôle in the calculation of the visibility of the new moon, so essential for the oriental moon calendar.

The interest in these magnitudes α_1 to α_{12} lies furthermore in their relation to the determination of the variability of the length of the days during the seasons. This is an immediate consequence of the following consideration. From the fact that ecliptic and horizon are both great-circles on the celestial sphere it follows that, at any moment, one half of the ecliptic is above the horizon,[12] or, in other words, six of the zodiacal signs. On the other hand, the length of a day is the time from sunrise to sunset during which period six signs rose one after the other. These signs compose the semicircle above the horizon at the moment of sunset. This gives the fundamental theorem: *The length of a day corresponding to a certain position of the sun in the ecliptic equals the sum of the rising times for the next 180 degrees of the zodiac.*

In order to express this relation by a short formula, we introduce twelve magnitudes C_1, C_2, $\cdots$, C_{12} by the following definition: C_1 gives the length of the day corresponding to the sun standing in the vernal point, C_2 the length of the day corresponding to a sun position 30 degrees more advanced and so on to C_{12} when the sun stands at the beginning of the twelfth sign. The theorem which connects the ἀναφοραί and the length of the days is then simply

$$C_1 = \alpha_1 + \alpha_2 + \cdots + \alpha_6$$
$$C_2 = \alpha_2 + \alpha_3 + \cdots + \alpha_7$$

(1)

$$C_{12} = \alpha_{12} + \alpha_1 + \cdots + \alpha_5$$

These relations are essential for the understanding of all ancient discussions of the rising times of the zodiacal signs.

A second system of formulae is equally important, establishing the symmetry of the rising times with respect to the vernal point and the autumn point, namely

(2)
$$\begin{aligned}\alpha_1 &= \alpha_{12} & \alpha_4 &= \alpha_9 \\ \alpha_2 &= \alpha_{11} & \alpha_5 &= \alpha_8 \\ \alpha_3 &= \alpha_{10} & \alpha_6 &= \alpha_7\end{aligned}$$

as can be seen by simple consideration of a sphere. Both relations (1) and (2) are general facts independent of the actual values of the α's.

6. The problem of determining the values of the α's can obviously be considered as a problem of spherical trigonometry and was undoubtedly one of the main causes of the development of this field. We find its complete solution by trigonometrical methods given in Ptolemy's "Almagest" (about 150 A.D.) but it is significant for the importance attributed to this question that one half of Ptolemy's "Planisphaerium"[13] is devoted to another solution by such absolutely different methods as stereographic projection and considerations which we today would call descriptive geometry and nomography. From earlier sources, however, only approximative solutions are testified and these methods we intend to discuss here.[14] The essential feature in this preliminary stage in the determination of the rising times is the assumption that the values of the α's increase or decrease, respectively, by a constant amount d. We call these methods therefore "*linear methods*," in contrast to the exact "trigonometrical methods."[15]

These "linear" methods appear in two slightly different forms. The "strictly" linear method assumes that the increase of the α's from α_1 to α_6 is always d or

(3a)
$$\begin{aligned}\alpha_2 &= \alpha_1 + d = \alpha_{11} & \alpha_5 &= \alpha_4 + d = \alpha_8 \\ \alpha_3 &= \alpha_2 + d = \alpha_{10} & \alpha_6 &= \alpha_5 + d = \alpha_7 \\ \alpha_4 &= \alpha_3 + d = \alpha_9\end{aligned}$$

The "generalized" linear method, however, assumes in the middle of the α-sequence twice as much as elsewhere or

(3b)
$$\begin{aligned}\alpha_2 &= \alpha_1 + d = \alpha_{11} & \alpha_5 &= \alpha_4 + d = \alpha_8 \\ \alpha_3 &= \alpha_2 + d = \alpha_{10} & \alpha_6 &= \alpha_5 + d = \alpha_7 \\ \alpha_4 &= \alpha_3 + 2d = \alpha_9\end{aligned}$$

---

[12] This statement ignores not only atmospheric refraction but also the movement of the sun in the ecliptic during the day. For the approximative methods discussed in the following, both effects are of no influence.

[13] Ptolemy, opera II p. 225–259. As various remarks in this work prove, it was written after the Almagest (cf. e.g. p. 234, 16, p. 242, 2). It is usually assumed that this method had been invented by Hipparchus; cf. Delambre HAA II p. 454 f. The main source is a letter of Synesius (Migne PG 66 p. 1577, transl. Fitzgerald p. 263); the passage from Proclus, quoted by Delambre II p. 454, seems to be spurious.

[14] The development of the "trigonometrical methods" in Greek astronomy will be discussed in the thesis of Mr. Olaf Schmidt, Brown University.

[15] Ptolemy, Tetrabiblos I, 20 (ed. Robbins p. 94/95) = I, 21 (ed. Boll-Boer p. 46) speaks about "the common method, based upon evenly progressing increases in the ascensions which is not even close to the truth" (transl. Robbins).

In the following, we refer to these two kinds of assumption as "System A" (3a) and "System B" (3b) respectively.

7. The problem of determining the length of day and night from month to month according to the position of the sun in the zodiac is reduced by means of the relations (1) to the determination of the twelve magnitudes α_1 to α_{12}. The symmetry-relation (2) reduces the number of unknown quantities to six. If we furthermore make the assumption that the variation of the α's follows either the scheme A or B, the problem is still more simplified since because of (3) only two quantities, say α_1 and d, remain unknown. Therefore, we need only two independent relations in order to know the length of day and night for any given position of the sun.

Two such relations are furnished by the knowledge of the length M of the longest day and the length m of the shortest day. M occurs when the sun's distance from the vernal point amounts to 90°, i.e. M is the same as C_4. Hence, from (1) and (2) follows

$$(4') \qquad M = 2(\alpha_4 + \alpha_5 + \alpha_6)$$

and correspondingly

$$(4'') \qquad m = 2(\alpha_1 + \alpha_2 + \alpha_3).$$

If we work with "System A," then (3a) gives

$$(5a) \qquad \begin{aligned} M &= 6\alpha_1 + 24d \\ m &= 6\alpha_1 + 6d \end{aligned}$$

and (3b) gives correspondingly in the case of "System B"

$$(5b) \qquad \begin{aligned} M &= 6\alpha_1 + 30d \\ m &= 6\alpha_1 + 6d. \end{aligned}$$

From (5) both α_1 and d can be easily calculated and this gives us all the α's and C's.

This shows that if we know, for a given place, the length M of its longest day and the length m of its shortest day, then we are able to calculate all α's, from which results the law of the change of the length of the days during the year. Ancient astronomy simplified the problem further by assuming that the length m of the shortest day equals the length of the shortest night, or by assuming $M + m = 24$ hours.[16] The knowledge of M alone is therefore sufficient to characterize the variation of the days during the seasons.[17]

The essential point in this argumentation is the assumption of the law of variation of the α's, according to either system A or B. Both these systems are only approximations of the real law of dependence of the rising times on the sun's longitude, but it is necessary to emphasize that the obtained results are quite satis-

factory as far as the resulting values of the lengths of the days are concerned, at least if we take the inaccuracy of ancient time-measurement into consideration.

8. In order to give concrete examples of the preceding general discussion, we shall, at the end of this paragraph, give the values for the α's and C's as derived from cuneiform sources. However, we must first introduce some remarks concerning the units used to measure time.

The fundamental unit is the Sumerian length measure "danna" (called *bēru* in Akkadian), twelve of which correspond to one day's travel.[18] "Danna" usually is translated by "double hour," but without real reason, because this word does not contain any such element as "double" or "hour." Perhaps, the correct correspondence would be "mile." One sixtieth of this unit is called "stadium" by Manilius.[19]

These "miles" are subdivided into 30 parts each, called "uš," i.e. "length."[20] The transfer of these terrestrial units to the sky, according to the rule that twelve danna equal one day, thus creates the subdivison of the celestial equator into $12 \cdot 30 = 360$ parts. The "uš" corresponds therefore to our "degree," Greek $\mu o \hat{\iota} \rho a$.[21] In order to unify our notation, we shall express in the following both the α's and the C's consistently in "degrees," whether or not the original source uses degrees or (equinoctial) hours. Babylonian sources never use any other unit, but in Greek astronomy both units appear. The rule of transformation is for instance given in Mich. Pap. III 149:[22] $\mathit{ἰσχύουσιν δὲ οἱ \overline{ιε} χρόνοι \bar{α} ὥραν ἰσημερίνην.}$

We now are prepared to give the values for the $\mathit{ἀναφοραί}$ in Babylonian astronomy of Seleucid times. System A is based on the following values:

$$(6a) \quad \begin{aligned} \alpha_1 &= \alpha_{12} = 20° & \alpha_4 &= \alpha_9 = 32 \\ \alpha_2 &= \alpha_{11} = 24 & \alpha_5 &= \alpha_8 = 36 \quad d = 4 \\ \alpha_3 &= \alpha_{10} = 28 & \alpha_6 &= \alpha_7 = 40 \end{aligned}$$

from which the following lengths of the days result

---

[16] This is only correct if we again disregard atmospheric refraction which makes m greater than the complement to M.

[17] For explicit formulae, expressing d and α_1 by M see p. 255.

[18] In older literature, names such as KAS–BU and similar words can be found, because the sign "danna" contains two components which can be read as KAS and BU respectively.

[19] Manilius III, 282 ff.

[20] Here too, many misreadings are current in literature, for example geš or *imdu*. Even a non-existing Greek root $\mathit{στάω}$ has been introduced in order to explain a wrong Akkadian etymology. Cf. Neugebauer [2] p. 274 note 126.

[21] Hypsicles (I quote according to the text established by V. de Falco for a new edition of the "Anaphorikos" under preparation): Τοῦ τῶν ζῳδίων κύκλου εἰς $\overline{τξ}$ περιφερείας ἴσας διῃρημένου, ἑκάστη τῶν περιφερειῶν μοῖρα τοπικὴ καλείσθω· ὁμοίως δὴ καὶ τοῦ χρόνου, ἐν ᾧ ὁ ζῳδιακὸς ἀφ' οὗ ἔτυχε σημείου ἐπὶ τὸ αὐτὸ σημεῖον παραγίγνεται, εἰς $\overline{τξ}$ χρόνους ἴσους διῃρημένου, ἕκαστος τῶν χρόνων μοῖρα χρονικὴ καλείσθω. (Cf. ed. Manitius col. 5,25 ff.). Analogously Ptolemy, opera II p. 74,6 ff. ($προθ. τῶν πλαν. 3$): διαιρεθείσης δὲ τῆς περιφερείας αὐτοῦ (sc. of the equator) εἰς ἴσα τμήματα $\overline{τξ}$ καλείσθω τὰ τμήματα ἰδίως χρόνοι.

[22] XII, 9 f.

according to (1):

$$
\text{(7a)} \quad
\begin{array}{ll}
C_1 = 3{,}0° & C_7 = 3{,}0 \\
C_2 = 3{,}20 & C_8 = 2{,}40 \\
C_3 = 3{,}32 & C_9 = 2{,}28 \\
C_4 = M = 3{,}36 & C_{10} = m = 2{,}24 \\
C_5 = 3{,}32 & C_{11} = 2{,}28 \\
C_6 = 3{,}20 & C_{12} = 2{,}40.
\end{array}
$$

System B assumes

$$
\text{(6b)} \quad
\begin{array}{llll}
\alpha_1 = \alpha_{12} = 21° & \alpha_4 = \alpha_9 = 33 & & \\
\alpha_2 = \alpha_{11} = 24 & \alpha_5 = \alpha_8 = 36 & d = 3 & 2d = 6 \\
\alpha_3 = \alpha_{10} = 27 & \alpha_6 = \alpha_7 = 39 & &
\end{array}
$$

and thus

$$
\text{(7b)} \quad
\begin{array}{ll}
C_1 = 3{,}0° & C_7 = 3{,}0 \\
C_2 = 3{,}18 & C_8 = 2{,}42 \\
C_3 = 3{,}30 & C_9 = 2{,}30 \\
C_4 = M = 3{,}36 & C_{10} = m = 2{,}24 \\
C_5 = 3{,}30 & C_{11} = 2{,}30 \\
C_6 = 3{,}18 & C_{12} = 2{,}42.
\end{array}
$$

For sun positions at arbitrary points between the boundaries of the zodiacal signs, linear interpolation is adopted. This holds for all the lists discussed in the following.

The values C in (7a) and (7b) were discovered by Father Kugler in cuneiform tablets concerning the calculation of the new and full moons.[23] He established at the same time that system B is of later date than system A,[24] a statement which has been supported by recent investigations.[25]

As stated above, knowledge of the magnitude M is sufficient to determine all the α's, after the single assumption of the "system" to be used. If we express all units in degrees, then $M + m$ becomes $6{,}0°$ and from (5) p. 254 it follows that

$$
\text{(8)} \quad d =
\begin{cases}
\dfrac{1}{9}(M - 3{,}0) & \text{system A} \\[2ex]
\dfrac{1}{12}(M - 3{,}0) & \text{system B}
\end{cases}
$$

and

$$
\text{(9)} \quad \alpha_1 =
\begin{cases}
1{,}20 - \dfrac{5}{18}M & \text{system A} \\[2ex]
1{,}15 - \dfrac{1}{4}M & \text{system B}
\end{cases}
$$

from which all further α's can be calculated.

§ 3. The Anaphorai in Pap. Mich. III, 149

9. The two Babylonian schemes for the length of the days, as given in the preceding section, agree as to the extremal values $M = 3{,}36$ and $m = 2{,}24$; that is, they have the ratio $M : m = 3 : 2$ in common. This shows that they both refer to the same geographical latitude. The investigation of astronomical theories contained in the cuneiform tablets of Seleucid times does not reveal that Babylonian astronomy extended the scheme for the C's and α's to different latitudes. Until proof of the contrary, we may therefore assume that the idea of introducing variable geographical latitudes originated in Greek astronomy.

The word "latitude," however, must not be taken in the strict modern sense of one of the spherical coordinates on the earth. The geographical latitude does not occur in the Greek tables of obliquascensional times until they reached their final form, based on spherical trigonometry, as given in Ptolemy's Almagest, book II, chapter 8. But even these tables are not compiled with respect to linear increase in geographical latitude, but with respect to the maximum length of the day (our M), which increases in Ptolemy's tables by the constant amount of one half hour.[26] In the Geography we are instructed to draw the παραλλήλοι in such a way that M increases from the equator to the 14th parallel ($\varphi = 45$) by the constant amount of $\frac{1}{4}$ of an hour, from the 15th to the 19th by $\frac{1}{2}$ hour and by one hour between the three last ones.[27] This obviously reflects the older use of the non-trigonometrical or "linear" tables which we now proceed to consider.

The Mich. Pap. III, 149 contains the rising times for Aries and Libra (i.e. α_1 and α_6 in our notation) for seven different locations,[28] the so-called "seven climata." The content of the text[29] can be condensed in the following list:

[23] Kugler [1].

[24] Unfortunately called by him in reversed order "System II" and "System I" respectively. A more exact date of the origin than "in the beginning of the Seleucid period" cannot be given. Cf. p. 262 note 61.

[25] Neugebauer [2]. A third scheme has been proposed by Kugler, but is undoubtedly wrong, as Schnabel [1] p. 32 f. remarked and as can be proved by consequences in the calculation of the moon's crescent. Honigmann's discussion M. P. III p. 317 therefore has no basis whatsoever.

[26] Seven climata in Almagest II, 13 and VI, 11 (opera I p. 174 ff. and p. 538/9) and in the Analemma (opera II p. 282, 17 ff. cf. also p. 160, 2 ff.), only five in the φάσεις (opera II p. 4).

[27] Ptolemy, Geogr. I, 23. For a modern treatment cf. Mžik [1].

[28] In the preceding section I propose the following restoration in XI, 32: αϑ[η]ορη[τος].

[29] M. P. III, 149, col. XI, 40 to XII, 6.

climata	1 Ethiopia	2 Syria	3 Rhodes	4 Asia, Ionia	5 Argos	6 Rome, Italy, Marit. Gaul	7 [Asia], Germany, Britain [30]
(10) α_1	22	21	20	19	18	17	16
α_6	38	39	40	41	42	43	44
d	2;40	3	3;20	3;40	4	4;20	4;40

The numbers d are mentioned as the "προσθέσεις" or "ἀφαιρέσεις" (amounts to be added or subtracted) in order to complete the table for each clima. Unfortunately, there is one more sentence [31] ἐν παντὶ οὖν κλίματι ὁ κάρκινος καὶ ὁ αἰγόκερος ἐν λ χρόνοις ἀνέχθησαν i.e. for all latitudes $\alpha_4 = \alpha_{10} = 30°$. The editors therefore reconstructed the tables for the α's in such a way that $\alpha_4 = \alpha_{10}$. For example, for the second clima this resulted in the following table

(11)
$$\begin{aligned}
\alpha_1 &= 21 \\
\alpha_2 &= 24 = \alpha_{12} \\
\alpha_3 &= 27 = \alpha_{11} \\
\alpha_4 &= 30 = \alpha_{10} \\
\alpha_5 &= 33 = \alpha_9 \\
\alpha_6 &= 36 = \alpha_8 \\
\alpha_7 &= 39
\end{aligned}$$

and correspondingly for the other climata. These tables preserve the extreme values and the differences as given by the text but differ from the text by assuming that the maximum value refers to α_7 instead of to α_6. Not only do these tables not fulfill the fundamental relations (2), but also the corresponding lengths of the days become impossible. For example, the "equinoctial" day should be, according to definition,

$$C_1 = C_7 = 3{,}0 = \alpha_1 + \alpha_2 + \cdots + \alpha_6.$$

Using the above assumed numbers, it is only 2,52, and the resulting "equinox" has unequal length of day and night!

There seems to me to be no reason for assuming that such obvious contradictions should have escaped any ancient astronomer. Therefore we must construct the tables from the numbers given and thereafter discuss the meaning of the sentence in question. If we proceed in this way, we can easily give a perfectly correct scheme for the α's in seven climata by use of the elements (10) given in the papyrus. We need only remark that the given differences, together with the values of α_1 and α_6, make it necessary to insert $2d$ somewhere, or in other words, to use system B instead of system A. This gives the following scheme:

	climata	1	2	3	4	5	6	7
	$\alpha_1 = \alpha_{12}$	22	21	20	19	18	17	16
	$\alpha_2 = \alpha_{11}$	22;40	24	23;20	22;40	22	21;20	20;40
	$\alpha_3 = \alpha_{10}$	27;20	27	26;40	26;20	26	25;40	25;20
(12)	$\alpha_4 = \alpha_9$	32;40	33	33;20	33;40	34	34;20	34;40
	$\alpha_5 = \alpha_8$	35;20	36	36;40	37;20	38	38;40	39;20
	$\alpha_6 = \alpha_7$	38	39	40	41	42	43	44
	d	2;40	3	3;20	3;40	4	4;20	4;40

The numbers of the second clima are identical with the numbers already known from (6b) p. 255. This arrangement not only makes the assumption of serious errors unnecessary, but also reveals a perfectly clear structure: Deriving the C's from it in the usual way, we obtain for the shortest and longest days:

	climata	1	2	3	4	5	6	7
(13)	m	2,28	2,24	2,20	2,16	2,12	2,8	2,4
	M	3,32	3,36	3,40	3,44	3,48	3,52	3,56
	$M : m$	$\dfrac{53}{37}$	$\dfrac{3}{2}$	$\dfrac{11}{7}$	$\dfrac{28}{17}$	$\dfrac{19}{11}$	$\dfrac{29}{16}$	$\dfrac{59}{31}$

---

[30] The restoration Ασ[ιαι] in XII, 5 seems to me the only possible one in spite of the larger space available (ca. 5 letters destroyed) and the appearance of "Asia" in the fourth column.

[31] M. P. III, 149, XII, 7 ff.

This shows that the climata are arranged in arithmetical progression for m and M. Furthermore, the second clima "Syria" is distinguished by having the simple relation $M : m = 3 : 2$ exactly as Babylon. In other words: the list (12) takes the second clima as being equivalent to Babylon and starting from this latitude, extends the scheme to north and south in such a way that the longest days vary in arithmetic progression.

This scheme is in itself so simple and convincing that I have no doubt that it represents the right interpretation of the text. We can, however, deduce still more arguments from the text itself to show that system B and the latitude Babylon–Syria constitute the basis for this list. The papyrus mentions in XI, 28 ff. that the vernal point ought to be placed at the eighth degree of Cancer and this is exactly the same definition as in the later Babylonian theory ("B"). Furthermore, it is said in XII, 22 ff. and XII, 32 that the day is 14 hours when the sun is in Gemini. According to (1), we get for this magnitude and for the second clima from (12)

$$C_3 = \alpha_3 + \alpha_4 + \cdots + \alpha_8 = 3,30° = 14^h$$

exactly as the text. Assuming (11), one would obtain $C_3 = 3,21° = 13;24^h$. For the sun in Taurus the text gives in XII, 20 the length of the day as 13 hours, where (12) results in $C_2 = 3,18° = 13;12^h$ (cf. p. 253), but (11) in $C_2 = 3,9° = 12;36^h$. In the following, we shall meet different instances where fractional parts are simply disregarded; then again 13^h is the result only from (12) but not from (11).

It remains to return once more to the questionable sentence that "in every clima Cancer and Capricornus ascend during 30 χρόνοι." We have, I think, two possibilities: either we consider this sentence as the erroneous addition of some scribe, or we take into consideration that the papyrus puts the vernal point at the eighth degree of Cancer. Under this assumption, the arcs of 30 degrees, to which the α's refer, overlap the zodiacal "signs." "Cancer" belongs then partly to α_3, partly to α_4 and "Capricornus" both to α_9 and α_{10}. If we now look at the scheme (12), one could say that a sign which lies "between" α_3 and α_4 or α_9 and α_{10} corresponds for every climata to the average value of 30°, represented in (12) by the dotted line. The sentence in question would therefore be fully correct if the vernal point lay in the 15th degree of the first "sign," but also remains at least intelligible as a loose expression under the present conditions. This explanation remains, of course, open to discussion, but it is certainly unnecessary to derive a contradictory scheme from this single casual remark.

10. Having established the scheme (12) as the method of determining the α's and C's according to system B and Babylon as the basic latitude, we now shall show that this method is in perfect agreement with other Greek sources.

System B is applied by Cleomedes [32] in his law for the variability of the lengths of the days at the Hellespontic clima,[33] which is characterized by $M : m = 5 : 3$.[34] The same rule is given by Martianus Capella [35] (end of 4th cent.) and Gerbert [33] (the pope Silvester II, end of 9th cent.). The resulting α's are

$$(14) \quad \begin{array}{l} \alpha_1 = \alpha_{12} = 18;45° = 1;15^h \\ \alpha_2 = \alpha_{11} = 22;30° = 1;30 \\ \alpha_3 = \alpha_{10} = 26;15° = 1;45 \\ \text{-----------------} \quad d = 3;45° = 0;15^h \\ \alpha_4 = \alpha_9 = 33;45° = 2;15 \\ \alpha_5 = \alpha_8 = 37;30° = 2;30 \\ \alpha_6 = \alpha_7 = 41;15° = 2;45 \,. \end{array}$$

System A is represented four times. The most important text is Vettius Valens. He first [37] gives the complete system of the α's exactly as in the Babylonian sources of system A (see p. 254 formula (6a) and (7a)). Then he gives the list of the "seven climata," [38] but now starting from Alexandria [39] (characterized by $M : m = 7 : 5$ [40]) and defining the "climata" such that the M's are in arithmetical progression. The rising times are:

[32] Cleomedes I, 6 ed. Ziegler p. 50.

[33] I believe to have shown that Cleomedes lived in Lysimachia on the Hellespont (Neugebauer [3]). These passages were interpreted as an arbitrary numerical example by Pogo [1] p. 413 f.

[34] That the longest day at the Hellespont was supposed to be 15 hours is well known; cf. e.g. Ptolemy, Almagest II, 8 (ed. Heiberg p. 138).

[35] Martianus Capella VIII,877 f. ed. Eyssenhardt p. 326 f., ed. Dick p. 462 f.

[36] Gerbert, opera, ed. Bubnow p. 39 f. (Gerbert omits one line).

[37] Vettius Valens I, 7 ed. Kroll p. 23.

[38] Vettius Valens I, 7 ed. Kroll p. 24, 8 ff.

[39] I do not see any "unlösbaren Widerspruch" as Honigmann does (M. P. III p. 302) in the fact that an author uses first one place (Babylon) for a special example of his method and thereafter starts the general list of the climata from another point (Alexandria). This reflects merely the historical facts that the method originated in and for Babylon only and that the scheme of the seven climata was a Hellenistic invention.

Bishop George, who lived around 700 A.D. in Mesopotamia, does exactly the same in giving the rising times according to system A for Babylon, the seven climata, however, according to the scheme of Almagest II, 13 (cf. above p. 255, note 26) which does not contain Babylon (Ryssel [1] p. 48 and p. 49).

[40] E.g. Hypsicles: ὑποκείσθω δὴ τὸ ἐν Ἀλεξανδρείᾳ τῇ πρὸς Αἰγυπτον κλίμα ἐν ᾧ ἡ μακροτάτη ἡμέρα πρὸς τὴν βραχυτάτην ἡμέραν λόγον ἔχει ὃν ξ πρὸς ε (cf. ed. Manitius col. 7, 3 ff.).

(15)

clima	1	2	3	4	5	6	7
$\alpha_1 = \alpha_{12}$	21;40	20;33,20	19;26,40	18;20	17;13,20	16; 6,40	15
$\alpha_2 = \alpha_{11}$	25	24;20	23;40	23	22;20	21;40	21
$\alpha_3 = \alpha_{10}$	28;20	28; 6,40	27;53,20	27;40	27;26,40	27;13,20	27
$\alpha_4 = \alpha_9$	31;40	31;53,20	32; 6,40	32;20	32;33,20	32;46,20	33
$\alpha_5 = \alpha_8$	35	35;40	36;20	37	37;40	38;20	39
$\alpha_6 = \alpha_7$	38;20	39;26,40	40;33,20	41;40	42;46,40	43;53,20	45
d	3;20	3;46,40	4;13,20	4;40	5; 6,40	5;33,20	6

from which follows

(16)

clima	1	2	3	4	5	6	7
m	2,30	2,26	2,22	2,28	2,14	2,10	2,6
M	3,30	3,34	3,38	3,42	3,46	3,50	3,54
$M:m$	$\dfrac{7}{5}$	$\dfrac{107}{73}$	$\dfrac{109}{71}$	$\dfrac{37}{23}$	$\dfrac{113}{67}$	$\dfrac{23}{13}$	$\dfrac{13}{7}$

The analogy between Vettius Valens and Mich. Pap. 149 is striking, the only differences being in systems A and B on the one hand, and in the selection of funda-mental latitude Alexandria and Babylon on the other. But also Vettius Valens calls the clima of Babylon the "second"[41] and both schemes define the climata by a constant difference of 4° as (13) p. 256 and (16) show, and as Vettius Valens explicitly states.[42]

System A for Babylon ($M : m = 3 : 2$) is once more followed by Manilius and by Bishop George,[43] for Alexandria ($M : m = 7 : 5$) in Hypsicles,[44] showing how familiar those "linear methods" were to Greek astronomers.

§ 4. Firmicus Maternus

11. In the second book of his "mathesis," Firmicus Maternus [45] gives the following list of rising times:

clima	Alexandria and Babylon	Rhodes	Hellespont	Athens	Ancona	urbs
$\alpha_1 = \alpha_{12}$	20	19	17	18	15	17
$\alpha_2 = \alpha_{11}$	24	23	22	23	21	22
$\alpha_3 = \alpha_{10}$	28	27	27	27	27	27
$\alpha_4 = \alpha_9$	32	32	32	32	32	32
$\alpha_5 = \alpha_8$	36	36	37	36	38	37
$\alpha_6 = \alpha_7$	40	40	42	41	44	42

The disorder is obvious, except in the first clima which is identical with the scheme A for Babylon. Honigmann recognized correctly [46] that the numbers can be explained by the assumption that the fractional parts are ignored, and he restored the following original list:

clima	[I]	II	III	IV	V	[VI]	VII
$\alpha_1 = \alpha_{12}$	21;40	20	19;¦26,40	18;¦53,20	17;¦30	16;40	15
$\alpha_2 = \alpha_{11}$	25	24	23;¦40	23;¦20	22;¦30	22	21
$\alpha_3 = \alpha_{10}$	28;20	28	27;¦53,20	27;¦46,40	27;¦30	27;20	27
$\alpha_4 = \alpha_9$	31;40	32	32;¦ 6,40	32;¦13,20	32;¦30	32;40	33(!)
$\alpha_5 = \alpha_8$	35	36	36;¦20	36;¦40	37;¦30	38	39(!)
$\alpha_6 = \alpha_7$	38;20	40	40;¦33,20	41;¦ 6,40	42;¦30	43;20	45(!)
M	3,30	3,36	3,38	3,40	3,45	3,48	3,54

---

[41] Vettius Valens I, 14 ed. Kroll p. 28, 24.   [42] Vettius Valens I, 7 ed. Kroll p. 24, 17 f.
[43] Manilius Astron. III, 275 ff.; Ryssel [1] p. 47 f.
[44] Hypsicles, Anaphorikos, cf. note 40. Quoted also by Ptolemy, Tetrabiblos, I, 20 (ed. Robbins p. 94/95) = I, 21 (ed. Boll-Boer p. 46).   [45] Firmicus Maternus II, 11 ed. Kroll-Skutsch p. 53 ff.   [46] Honigmann [1] p. 45.

He had to assume the omission of [I] and [VI], the duplication of V and three errors of one unit each in VII. Moreover, this scheme has not the fundamental property of linear increase of maximal lengths of the days from clima to clima as found in Vettius Valens and Pap. Mich. III, 149.

There exists however, still another possibility, which, omitting the fractional parts, gives the numbers of Firmicus exactly without any correction and explains the duplication of the column beginning with 17. We saw that Vettius Valens started his list from Alexandria, increasing the value of M by 4 from clima to clima; on the other hand, Pap. Mich. III, 149 began from Babylon, but with the same difference in M (and using system B instead of A). The sets of M-values subdivide each other symmetrically, as the following comparison shows

(a) Vettius Valens: 3,30   3,34   3,38   3,42   3,46   3,50   3,54
(b) Mich. Pap. III, 149:   3,32   3,36   3,40   3,44   3,48   3,52   3,56

If we combine these two schemes (of course replacing system B in the second line by A) we get a scheme which contains the following tables:

climata	IIb	IIIa	IIIb	[IVa]	IVb	Va	[Vb]	[VIa]	VIb
$\alpha_1 = \alpha_{12}$	20	19;¦26,40	18;¦53,20	18;20	17;¦46,40	17;¦13,20	16;40	16; 6,40	15;¦33,20
$\alpha_2 = \alpha_{11}$	24	23;¦40	23;¦20	23	22;¦40	22;¦20	22	21;40	21;¦20
$\alpha_3 = \alpha_{10}$	28	27;¦53,20	27;¦46,40	27;40	27;¦33,20	27;¦26,40	27;20	27;13,20	27;¦ 6,40
$\alpha_4 = \alpha_9$	32	32;¦ 6,40	32;¦13,20	32;20	32;¦26,40	32;¦33,20	32;40	32;46,40	32;¦53,20
$\alpha_5 = \alpha_8$	36	36;¦20	36;¦40	37	37;¦20	37;¦40	38	38;20	38;¦40
$\alpha_6 = \alpha_7$	40	40;¦33,20	41;¦ 6,40	41;40	42;¦13,20	42;¦46,40	43;20	43;53,20	44;¦26,40
M	3,36	3,38	3,40	3,42	3,44	3,46	3,48	3,50	3,52

Except for the omission of three climata (and the restriction to integers) we have here exactly the numbers given by Firmicus Maternus. We do not even need to speak about "omissions." What Firmicus (or, better, his source) did was simply to take a list of rising times, progressing by 2° in maximal length of the days, and to select those groups which corresponded to seven famous cities.

12. Having achieved this insight into the simple structure of lists of rising times, i.e. for the purpose of arranging geographical zones according to linear increasing values of M, we now have to discuss the names associated with these zones. Obviously, much disorder prevails here, but it seems to me still possible to reach some understanding at least of the lists found in our sources. We saw that the leading principle in the order of all the tables of rising times is their arrangement according to increasing maximum length of the daytime. Taking this as granted, Firmicus' series of place-names becomes correct as far as the increase of latitude is concerned: Alexandria → Babylon → Rhodes → Athens → Hellespont → Rome → Ancona. The same is true in the case of Mich. Pap. III, 149 with the only exception of the totally misplaced "Argos":[47] Ethiopia → Syria → Rhodes → Asia and Ionia → Argos(!) → Rome, Italy, Gallia

maritima → As[ia], Germany, Britain. If we remove Argos and replace it by Rome, accepting Italy and Gallia as equivalents of Ancona, then agreement between Firmicus and the Mich. Pap. is reached from Babylon to Ancona.[48] The papyrus adds as southernmost clima Ethiopia, as northernmost Germany and Britain, doubtless incorrect as far as the latitudes are concerned, but obviously taken from some scheme considering these zones as boundaries of the oikumene.[49]

Serious difficulties occur if we consider the numbers in each table as exact representatives of the latitude of the locality heading the column. However, there exists no proof of the correctness of this assumption. We may just as well assume that the values given refer to the central line in a certain zone to which the place in question belongs, without lying exactly in the middle of the strip, the extent of each strip being determined simply by the regular increase of M from clima to clima.[50] The resulting arrangement is repre-

[47] Honigmann (M. P. III p. 304 note 10) assumes tentatively Argos Amphilochensis which in itself seems to be very unlikely and makes the arrangement no better than if we assume the Peloponnesian Argos.

[48] We therefore do not need to accept Honigmann's "überraschendes Ergebnis, dass die Benennungen der sieben Klimata in dem Papyros völlig aus der Luft gegriffen sind, mit ihnen nichts zu tun haben und von uns unbeachtet gelassen werden dürfen!" (M. P. III, p. 304).

[49] Ptolemy opera II p. 160, 2 ff. (προχ. καν.) speaks about τούτων μεταξύ πως τῆς οἰκησίμου παραλλήλων ἑπτὰ παρακειμένων. For Ethiopia on the one hand and the Ister on the other as boundaries of the oikumene see e.g. Heidel [1] p. 26 ff. and p. 31 ff.

[50] Diller [1] p. 264 came to the same result from absolutely different considerations.

M	3,30	3,35	3,40	3,45	3,50	3,55
cuneiform	Babyl.					
Hypsicles	Alexandr.					
Manilius	Babyl.					
Vettius Val.	I	II Babyl. III	IV	V	VI	VII
Mich. P.III, 149	Ethiopia (!)	Syria	Rhodes	Asia, Ionia	[Rome](!)	Italy, Gallia	Germany and Brit. (!)
Cleomedes			Hellespont			
Firmicus Mat.		Bab. Rhod. Athens		Hellesp. Rome		Ancona	
Almagest	Lower Egypt	Rhodes		Hellespont	Middle Pontus	
M	3,30	3,35	3,40	3,45	3,50	3,55

sented by the figure above. Except for the three above-mentioned errors in Mich. Pap. III, 149, no further correction of the tradition is necessary. The zones overlap each other in such a way that the same place names have always at least parts of the zones in common.

13. Honigmann drew attention to another source of information about the anaphorai,[51] namely the astrological calculation of the maximal possible length of the human life. This is the doctrine that the number of years can never exceed the maximal possible number κ of degrees which is necessary for one quarter of the ecliptic to rise. This number is obviously given by

$$\kappa = \alpha_5 + \alpha_6 + \alpha_7 = \alpha_6 + \alpha_7 + \alpha_8$$

and, according to Epigenes, has the value $112 = 1,52$; according to Berossos, $116 = 1,56$; and "in Italiae tractu," $124 = 2,4$.[52] Honigmann saw that the difference between these values must correspond to the difference in the respective clima and that Epigenes speaks about Alexandria, Berossos about Babylon. We can now formulate these results a little more precisely by calculating κ according to the two systems the existence of which we recognized in simultaneous use. We obtain for κ:

κ	A	B
Alexandria	1,51;40	1,50
Babylon	1,56	1,54

[51] M. P. III p. 307 ff. The necessary references are given there.
[52] Pliny N.H. VII, 160 (ed. Mayhoff II p. 56, 2).

This shows that the values given by Epigenes and Berossos are both based on system A. As to the value $\kappa = 2,4$ "in Italiae tractu," we find from Mich. Pap. III, 149 column 6 in (12) p. 256 the value 2,4;40 and this is just the column headed as "Rome, Italy, Gallia maritima." This shows again the correctness of the place names in the text.

§ 5. Pap. Mich. III, 151

14. A more elaborate type of obliquascensional table is preserved in Pap. Mich. III, 151, although in such a fragmentary condition that I did not succeed in giving a complete restoration. The essential point is that this text gives the rising times not only from sign to sign, but from degree to degree, and furthermore, the first degree is subdivided into $\frac{1}{3}$ (= 0;20) and $\frac{2}{3}$ (= 0;40). The fragment contains parts of tables for five different zodiacal signs. The arrangement of these tables, however, is not such that each table

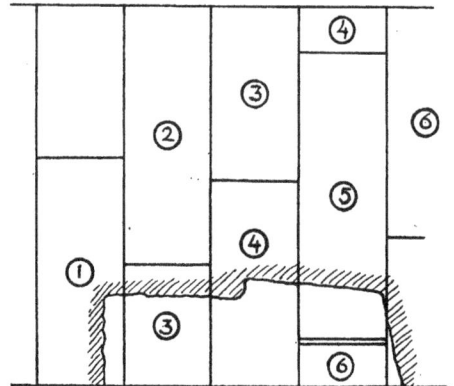

occupies one column, but table follows table as the height of the text permits. The previous figure explains this situation. A simple calculation shows that each column contained 46 lines from which only the lower quarter has been preserved.

In order to discuss the structure of these tables in detail, it is preferable to start with section 6, where the title ταυρου διδυμ[ου] is preserved. That two signs are mentioned may be explained by assuming the vernal point somewhere inside the sign Aries (e.g. 8° or 10°) such that α_2 belongs to both ♉ and ♊. The numbers given in the text are: [53]

section 6:	0,29,56	0,20
	0,59,52	0,40
	0,22,48	1
	2,59,36	2
	4,29	3
	5,59	4

If we take 0;29,56 as ⅓ of some unit, as indicated by the numbers in the second column, we obtain

0;29,56	0;20
0;59,52	0;40
1;29,48	1
2;59,36	2
4;29,24	3
5;59,12	4

This indicates that in line 3 the papyrus has the erroneous figure 0,22 instead of 1,29 and that the last place in the last two lines is omitted. That this omission was intentional is shown by section 3: [54]

section 3:   calculation:

1.	[0,27 ,10]	[0;20]		0;27,10	0;20
	[0,54 ,20]	[0;40]		0;54,10	0;40
	1,21[,30]	1		1;21,30	1
	2,43	2		2;43	2
5.	4,30	3		4; 4,30	3
	9,26	4		5;26	4
	6,47	5		6;47,30	5
	8, 9	6		8; 9	6
	10,52	8		9;30,30	7
10.	12, 3	9		10;52	8
	13,35	10		12;13,30	9
	14,56	11		13;35	10
				14;56,30	11

Here, the text omits all 30's except in line 5, where 4,30 is an error for 4,4,30. [55]

We are now able to restore section 4. Corresponding to 20 the text gives the number 30,36 and therefore

the number corresponding to 1 must be 30,36 : 20 = 1,31,48. We therefore should have

19	29, 4,12
20	30,36
21	32, 8,48

but the text gives 29,13 corresponding to 19 and 32,18 corresponding to 21. However, if we correct 30,36 to 30,46, we obtain full agreement, except, of course, for the omission of the third sexagesimal place: [56]

section 4:	1.	24,36	16	calculation:	24,36,48	16
		26, 9	17		26, 9, 6	17
		27,41	18		27,41,24	18
		29,13	19		29,13,42	19
	5.	30,36	20		30,46	20
		32,18	21		32,18,18	21
		33,50	22		33,50,36	22
		35,22	23		35,22,54	23
		35,55	24		36,55,12	24

15. Hence, we have reached a complete understanding of the arithmetical structure of sections 3, 4 and 6. Each section is an arithmetical progression δ, 2δ, 3δ, $\cdots$, 30δ, adding $\frac{1}{3}\delta$ and $\frac{2}{3}\delta$ at the beginning, and in general omitting the third place of all numbers. Each section is therefore completely characterized by one number, say 30δ, as follows:

section:	3	4	6
30δ:	40;45	46;9	44;54

We know, furthermore, that section 6 refers to the signs ♉ and ♊, and it is obvious that the numbers $1, 2, \cdots, 30$ refer to degrees. Such lists would be the analogue of the κανόνιον τῶν κατὰ δεκαμοιρίαν ἐπισυναγόμενοι in Ptolemy Almagest II, 8 giving the "χρόνοι ἐπισυναγόμενοι" but from degree to degree. The only objection to this explanation of our tables seems to be the magnitude of the numbers involved: rising-times of more than 40 degrees are possible only in the northernmost climata and then never for signs around ♉. This can however easily be explained by the choice of units. As mentioned above p. 254 Manilius expresses the obliquascensional times in "stadia" and one stadium equals two degrees. If we assume the same units to be applied here, then the right order of magnitude is restored. 44;54 stadia = 22;27° is a perfectly possible value for α_2 (e.g. in the fifth clima $\alpha_2 = 22°$ in Mich. Pap. III, 149 and $\alpha_2 = 22;20$ in Vettius Valens and $\alpha_2 = 22;30°$ on the Hellespont according to Cleomedes). [57]

We must still consider section 5. Two lines only are preserved: [58]

52,45	29
54,13	30

---

[53] Line 4: β νθ λς instead of β νδ λς as given in the edition according to friendly investigation of the original by Dr. F. E. Robbins.

[54] Line 3: ᾳ ϙᾳ Robbins.

[55] Real errors are: line 6: θ κγ instead of ε κγ. Omission of the line θ λ λ|ξ. Line 10: ιβ γ instead of ιβ ιγ.

[56] One more error in the last line: λε νε instead of λς νε.

[57] Cf. p. 257.

[58] Remainders from the preceding line: γ at the first place (Robbins).

and in the second line 54,13 could be replaced by 54,16 (i.e. γ by ς). However, at least one of these two lines must contain an error, because from $30\delta = 54$ follows $\delta = 1,48$, but the difference between 52,45 and 54,13 or 54,16 is only 1,28 or 1,31. A simple consideration of all possibilities of correcting this error shows that by far the most plausible assumption is the replacement of 40 by 50, leading to the following values in section 5:

$$1\delta = 1,28,27$$
$$29\delta = 42,45, 3$$
$$30\delta = 44,13,30$$

from which the text omits, as usual, the third sexagesimal place.

This finally would give us the following set of α's, contained in Mich. Pap. 151:

section	3	4	5	6
α	20;22,30	23;4,30	22;6,45	22;27

All these values are possible for α_1 and α_2 in the usually considered climata but I cannot find any relation which could define them exactly. The differences between these α's do not agree with the assumption of one geographical latitude. The only conclusion left, as far as I can see, is the following: sections 3 and 4 give the values for α_1 and α_2 for a clima about Rhodes. Sections 5 and 6 however both give α_2 but for two special places, about $1\frac{1}{2}$ or 2 degrees different in latitude, somewhere between Rhodes and the Hellespont.

§ 6. Conclusions

16. The representation of the anaphorai of the zodiacal signs by means of simple arithmetical series is an interesting example of how problems can be solved which seem to require trigonometrical methods. The anaphorai, however, are only one example of the treatment of periodic phenomena by linear approximations (or iterated linear approximations) which was developed to a surprising degree by Babylonian astronomy of Seleucid times and has gained the highest admiration of every scholar who studied this subject.

Having adopted the type of approximation, either A or B as previously designated, the anaphorai for all zodiacal signs are determined by one single constant M, the length of the longest day, or by the equivalent ratio $M : m$ assumed to be $3 : 2$ in Babylon. It is interesting to realize that this method was not merely adopted by the Greeks for special latitudes like Babylon or Alexandria, but consistently expanded as soon as the discovery of the sphericity of the earth required the consideration of changing horizons.[59] The Greek definition of the climata by a linear sequence of values

of M instead of by the geographical latitude [60] shows exactly the same methodological idea as the Babylonian astronomy. These methods furnished the empirical material for the numerical determination of astronomically important magnitudes, and prepared the ground on which trigonometrical methods could be developed step by step. It is not surprising that these methods were even used long after the definitive creation of spherical trigonometry, because of their simplicity, not requiring elaborate tables.

It seems unnecessary to me, however, to assume the existence of any uniquely determined Greek doctrine of the climata, or a clear distinction between "astrological" and "astronomical" climata as Honigmann does. The preserved texts show the existence of tables of rising times (or equivalent tables of length of the days) for special places, like Alexandria, Babylon or the Hellespont, having in common only the choice of a convenient simple value for $M : m$. On the other hand, the tables for groups of climata (as Vettius Valens or Mich. Pap. III, 149) have no more in common than the general arrangement of the climata according to linear increasing values of M, but differ in the choice of the starting point (Babylon or Alexandria) and in the system adopted (A or B).

17. As stated previously, the Babylonian origin of the "linear" theory of the anaphorai is obvious. More difficult is the determination of the time of adoption of the basic ideas. The Babylonian methods of calculating the movement of the sun and the moon and the planets from which we have our information about the two systems A and B cannot be traced much further back than 200 B.C. The oldest dated tablet, although belonging to the later system (B), refers to the year 208 B.C.[61] Our present source material, very complete for the period from 208 B.C. to 46 B.C., does not require the assumption of a date for the highest level of Babylonian theoretical astronomy earlier than, say, 250 B.C. We know, on the other hand, that the contact of Greek and Babylonian astronomy had its climax about 200 to 150 B.C.[62] This would fit in very well with the fact, pointed out especially by Honigmann, that the theory of the seven climata belongs to Eratosthenes.[63]

The restriction of the number of "climata" to seven marks neither the beginning nor the end of the

---

[59] That this happened later than in the fifth century, seems to me now definitely established by the researches of Frank ([1] p. 184 ff.) and Heidel ([1] p. 63 ff. esp. p. 79 f.).

[60] Cf. p. 255.

[61] The oldest text preserved is Chicago A 3430 + A 3431 (to be joined with Warka X 63, unpublished, No. 125 of my edition of cuneiform texts under preparation). Schnabel's attempts to date both systems ([1] p. 7 ff. and [2] p. 218 f. and p. 223 ff.) are based on wrong assumptions. Also Olmstead's attempt ([1] p. 122 f.; the quotation in note 30 is wrong; replace O. Schroeder by A. Ungnad) is inconclusive, because there is no proof that the Nabu-rimanni, who witnessed documents in 491/0, is the astronomer of system B whose name appears in a tablet of 46 B.C. (VAT 209 = No. 16 of my edition; published with many errors by Schnabel [2] p. 244 f.).

[62] Cumont [1].

[63] Honigmann [1] p. 10 ff. Cf. furthermore Diller [1] p. 263.

Greek concept of geographical zones. Ptolemy ignores the restriction to seven both in the φάσεις and in the Geography,[64] and the relationship of Epigenes' and Berossos' numbers of the longest possible lifetime to the anaphorai[65] supposes the existence of the Babylonian scheme "A" at least around 300 B.C.[66] This does not contradict our statement that Babylonian theoretical astronomy scarcely originated before 250 B.C. From Vettius Valens we know of attempts of undoubtedly Babylonian origin to determine the relation between the visibility of the new crescent and the anaphorai of the zodiacal signs.[67] This theory is much more primitive than the elaborate method known from moon-theory A. On the other hand, there exist much older cuneiform tablets which attempt to describe the relation of the moon's invisibility to the seasons by still rougher linear approximations.[68] This shows that the problem of the determination of the length of the days belongs to the oldest part of Babylonian astronomy, preceding the theory of the planetary movement and the theory of the moon of the Seleucid and Arsacid period in the same sense as these linear methods discussed here preceded the later Greek sphaeric.

Bibliography and Abbreviations

CLEOMEDES. Cleomedis, de motu circulari corporum caelestium, libri duo, ed. Ziegler. Leipzig, Teubner, 1891.

CUMONT [1]. F. Cumont, Comment les Grecs connurent les tables lunaires des Chaldéens, Florilegium de Vogüé. Paris, Geuthner, 1910.

DELAMBRE HAA. [J. B. J.] Delambre, Histoire de l'astronomie ancienne. Paris, 1817 (2 vols.).

DILLER [1]. A. Diller, Geographical latitudes in Eratosthenes, Hipparchus and Posidonius. Klio 27 (1934) 258–269.

DILLER [2]. A. Diller, The parallels on the Ptolemaic maps. Isis 33 (1940) 4–7.

FIRMICUS MATERNUS. Julii Firmici Materni matheseos libri VIII, ed. W. Kroll et F. Skutsch. Leipzig, Teubner, 1907, 1913.

FITZGERALD, SYNESIUS. Augustine Fitzgerald, The letters of Synesius of Cyrene. Oxford Univ. Press, 1926.

FRANK [1]. E. Frank, Plato und die sogenannten Pythagoreer. Halle, 1923.

GEMINUS. Geminus, Elementa astronomiae, ed. C. Manitius. Leipzig, Teubner, 1898.

GERBERT, opera. Gerberti postea Silvestri II papae opera matematica (972–1003), ed. N. Bubnov. Berlin, Friedländer, 1899.

HEIDEL [1]. W. A. Heidel, The frame of the ancient Greek maps. Am. Geogr. Soc., Research Series no. 20, New York, 1937

HONIGMANN [1]. E. Honigmann, Die sieben Klimata und die πόλεις ἐπίσημοι. Heidelberg, Winter, 1929.

HONIGMANN M. P. III. E. Honigmann, Die Anaphorai der alten Astrologen. M. P. III p. 299–321.

HYPSICLES. ANAPHORIKOS. Des Hypsicles Schrift Anaphorikos nach Überlieferung und Inhalt kritisch behandelt von K. Manitius, Programm d. Gymnas. z. heil. Kreuz. Dresden, 1888.

KUGLER [1]. F. X. Kugler, Babylonische Mondrechnung. Freiburg, Herder, 1900.

LANGE-NEUGEBAUER [1]. H. O. Lange-O. Neugebauer, Papyrus Carlsberg No. 1, Ein hieratisch-demotischer kosmologischer Text. Det Kgl. Danske Videnskabernes Selskab, Hist.-filol. Skrifter vol. 1 nr. 2 (1940).

MANILIUS, Astron. M. Manilius, Astronomica, ed. Th. Breiter. Leipzig, Dieterichs, 1908 and M. Manilii Astronomicon ed. A. E. Housman. London, Grant, 1903–1930.

MARTIANUS CAPELLA. De nuptiis philologiae et Mercurii ed. Eyssenhardt. Leipzig, Teubner 1866; ed. Dick, Leipzig, Teubner, 1925.

MIGNE PG. J.-P. Migne, Patrologiae cursus completus etc. Patrologia graeca.

M. P. III. Michigan Papyri vol. III, University of Michigan Studies. Humanistic Studies 40, Ann Arbor, 1936.

MŽIK [1]. H. v. Mžik, Des Klaudios Ptolemaios Einführung in die darstellende Erdkunde I. Wien, Gerold, 1935 = Klotho 5.

MŽIK [2]. H. v. Mžik, Review of Honigmann [1], OLZ 34 (1931) col. 939–945.

NEUGEBAUER [1]. O. Neugebauer, Some fundamental concepts in ancient astronomy. University of Pennsylvania Bicentennial conference. Studies in the history of science. Philadelphia, 1941, p. 13–29.

NEUGEBAUER [2]. O. Neugebauer, Untersuchungen zur antiken Astronomie III. QS Abt. B 4 (1938) p. 193–346.

NEUGEBAUER [3]. O. Neugebauer, Cleomedes and the meridian of Lysimachia. Am. J. Philol. 62 (1941) p. 344–347.

NEUGEBAUER-VOLTEN [1]. O. Neugebauer-A. Volten, Untersuchungen zur antiken Astronomie IV. Ein demotischer astronomischer Papyrus (Pap. Carlsberg 9). QS Abt. B vol. 4 p. 383–406 (1938).

OLMSTEAD [1]. A. T. Olmstead, Babylonian astronomy—historical sketch. AJSL 55 (1938) p. 113–129.

POGO [1]. A. Pogo, Egyptian water clocks. Isis 25 (1936), p. 403–425.

PTOLEMY Almagest. s. Ptolemy opera I.

PTOLEMY, Geography. Claudii Ptolemaei Geographia, ed. Nobbe. Leipzig, 1843.

PTOLEMY, opera. Claudii Ptolemaei opera quae extant omnia (Leipzig, Teubner). I. Syntaxis matematica (2 vols.) ed. Heiberg 1898, 1903. II. Opera astronomica minora, ed. Heiberg 1907. III, 1. Ἀποτελεσματικά ed. Boll et Boer 1940.

PTOLEMY, Tetrabiblos. ed. Robbins = Loeb Classical Library 1940. ed. Boll-Boer = Ptolemy, opera III, 1.

QS. Quellen u. Studien zur Geschichte der Mathematik, Astronomie und Physik.

ROBBINS [1]. F. E. Robbins, A new astrological treatise: Michigan papyrus no. 1. Classical Philology 22 (1927) p. 1–45.

RYSSEL [1]. V. Ryssel, Die astronomischen Briefe Georgs des Araberbischofs. ZA 8 (1893) 1–55.

SCHNABEL [1]. P. Schnabel, Kidenas, Hipparch und die Entdeckung der Praezession. ZA 37 (1927) p. 1–60.

SCHNABEL [2]. P. Schnabel, Berossos und die babylonisch-hellenistische Literatur. Leipzig, Teubner, 1923.

SETHE [1]. K. Sethe, Die Zeitrechnung der alten Aegypter im Verhältnis zu der der andern Völker, Nachr. d. K. Ges d. Wiss. zu Göttingen. Phil.-hist. Kl. 1919, p. 287–320; 1920, p. 28–55, 97–141.

SYNESIUS. s. Fitzgerald, Synesius and Migne PG 66, 1577 ff.

VETTIUS VALENS. Vettii Valentis anthologiarum libri, ed. W. Kroll. Berlin, Weidmann, 1908.

ZA. Zeitschrift für Assyriologie.

---

[64] Cf. note 49 p. 259 and note 26 p. 255.

[65] Cf. p. 260.

[66] Honigmann M. P. III p. 310 f.

[67] Vettius Valens I, 14; ed. Kroll p. 28.

[68] I intend to discuss this material from cuneiform sources separately in a forthcoming paper.

www.ingramcontent.com/pod-product-compliance
Lightning Source LLC
Chambersburg PA
CBHW081334190326
41458CB00018B/5991